James Logan Lobley

Hampstead Hill

Its Structure, Materials and Sculpturing

James Logan Lobley

Hampstead Hill
Its Structure, Materials and Sculpturing

ISBN/EAN: 9783337327217

Printed in Europe, USA, Canada, Australia, Japan

Cover: Foto ©berggeist007 / pixelio.de

More available books at **www.hansebooks.com**

Its Structure, Materials, and Sculpturing

BY

J. LOGAN LOBLEY, F.G.S., &c.,

PROFESSOR OF PHYSIOGRAPHY AND ASTRONOMY, CITY OF LONDON COLLEGE; AUTHOR OF
" MOUNT VESUVIUS," " GEOLOGY FOR ALL," ETC.

WITH

THE FLORA OF HAMPSTEAD,

BY

HENRY T. WHARTON, M.A., M.R.C.S., F.Z.S., &c.

AUTHOR OF " SAPPHO," ETC.

THE INSECT FAUNA OF HAMPSTEAD,

BY

THE REV. F. A. WALKER, D.D., F.L.S., F.E.S., F.R.G.S, &c

AUTHOR OF "L'ORIENT," " NINE HUNDRED MILES UP THE NILE," ETC.

AND

THE BIRDS OF HAMPSTEAD,

BY

J. EDMUND HARTING, F.L.S., F.Z.S. &c.

AUTHOR OF " THE BIRDS OF MIDDLESEX," ETC.

London:

ROPER AND DROWLEY, 11, LUDGATE HILL, E.C.

1889.

CONTENTS.

CONTENTS.

LIST OF ILLUSTRATIONS.

PREFACE.

THE recent great development of Hampstead as a residential locality has made the well-known Middlesex hill really a part of the metropolis. But although now joined to London, it still possesses many of those charms of Nature usually sought much further afield, and Hampstead Hill remains a happy hunting-ground for the naturalist.

That London has within its borders an area where the Botanist, the Entomologist, and the Ornithologist, as well as the Geologist, may profitably pursue their favourite studies in the field, is a remarkable fact; and to draw attention to this interesting feature of the great city, as well as to give the residents of Hampstead an increased pleasure and an additional interest in their delightful neighbourhood, is the object of the following pages.

The first section of this little book consists, with a few alterations and additions, of the matter of a series of articles that appeared in *The Hampstead and Highgate Express*, to the courteous proprietor of which my thanks are due for permission to reprint those articles, and for the use of some of the blocks required for the illustrations.

I have also to thank the Council of the Geologists' Association for allowing me to avail myself of the Geological Map of Hampstead which appeared in the third volume of the "Proceedings" of that Association, in illustration of a Paper by the late Mr. Caleb Evans, F.G.S.

It gives me much pleasure to acknowledge my great obligations to Dr. Wharton, the Rev. Dr. Walker, and Mr. J. E. Harting for very kindly furnishing me with their Lists of Hampstead Species, which derive a value both from not having been previously published and from the scientific position and long experience of their Authors.

J. L. L.

CITY OF LONDON COLLEGE,
     August, 1889.

HAMPSTEAD HILL.

CHAPTER I.

WHEN an English summer brings glorious English summer weather, which in some years, however, is slow in coming, the thoughts of Londoners instinctively turn to almost every place except their own wonderful town—to the sea-side, to the "Lakes," to Scotland, to Wales, to Killarney or the splendid coast of Wicklow, to the Rhine, to Switzerland—but still a few who are not altogether Philistines have visions not so far afield. To these there are sufficiently attractive charms in Greater London, that belt of greenery and red brick encircling the vast metropolis and wedding town to country with an emerald and ruby ring.

On every side of London this magnificent environment of the nation's capital is the wonder and delight of our north-country cousins, whose too often grimy towns and smoke-laden atmospheres seem as unfriendly to fine timber and luxuriant foliage as Nature is supposed to be to a vacuum. The wooded vales of Chislehurst, the heights of Norwood with their crystal crown, the Surrey commons, and the oak-studded glades of Richmond commanding the broad vale of

Thames rich with elm groves in *alto relievo*, the Castle Hill of Ealing, the classic hill of Harrow, the minor but no less beautiful Wembly and Dollis Hills, with Kingsbury's secluded fields and ancient church and the finest lake in the south of England lying between, all give to denizens of the great city of the Thames spots of beauty close at home. Another sweep of the wide circle holds in its embrace the grand old glades of Greenwich and of Epping Forest, the green meadows and slopes of Tottenham, Southgate, and Hornsey, with the spire-crowned hill of Highgate sparkling with its mansions of the olden time. But that which most adorns this splendid zone of light and life is picturesque Hampstead. It connects the two great sweeps of the circle and completes the whole ring with a brilliant and worthy gem.

How lovable and enjoyable and comforting to the spirit is such English scenery as that which encompasseth the great city. Not so the cold and awful Alpine peaks of rock and snow, nor the desolate and barren summits of our own mountain regions. These are by all means to be seen and explored, for the impressions they leave on the mind are deep and soul-elevating, but they are scarcely scenes to pass a life-time amongst, at least by Englishmen, who, doubtless, mentally moulded by their habitual environment, are most content where verdant vales and wooded gentle slopes delight the eye and assure the mind of peaceful repose and comfortable abundance. And who can see Hampstead itself without admiring the neighbourly contiguity of its houses and their picturesque groupings and settings, the opulent appearance and beauty of its outer villas and mansions, and the luxuriant foliage with which they are surrounded, and, above all, the great open Heath, which, from its commanding position and elevation, gives on one side a view of the town-filled valley of the Thames, and, on the other, a magnificent prospect of a perfect country landscape. Well may the inhabitants be proud of their town and its noble Heath.

The Heath, indeed, has a national importance, for it has long been a place of healthful and invigorating recreation for the people of London. To compare small things with great, the hill mass of Hampstead with its heath-crowned summit is as markedly prominent in Middlesex as is Switzerland in Continental Europe, and, as a consequence, it may be said that as that land of mountains has become the "playground of Europe" in like manner is Hampstead, so to speak, the playground of the English metropolitan district. But how few of the multitudes of people, who every year visit Hampstead and enjoy its breezy Heath, remember that their pleasure and the beauty and characteristics of the place and its surroundings are due to the internal structure of the hill, and the character of the ground below their feet.

MAP TO ILLUSTRATE THE GEOLOGY OF HAMPSTEAD.

A.—Bagshot Sands.
B.—Upper Sandy London Clay.
C.—London Clay (Main Bed).

1.—Jack Straw's Castle.
2.—Hampstead Heath.
3.—Well, Lower Heath.
4.—Child's Hill.
5.—Burgess Hill.
6.—Well Walk.
7.—Conduit Spring.
8.—Belsize Park.
9.—Shaft, North London Railway.
10.—West Heath.
11.—Flagstaff.
12.—Leg of Mutton Pond.
13.—Church Row.
14.—Vale of Health.
15.—Hampstead Ponds.
16.—North End.
17.—Platt's Lane.
18.—West End.
19.—Frognal House.
20.—Oakhill Park.
21.—Parish Church.
22.—Shaft, Midland Railway.
23.—Haverstock Hill.
24.—Finchley Road.
25.—Brick fields.
26.—Kidderpore Hall.
27.—High Street.
28.—Roslyn Bank.
29.—Flask Walk.
30.—Downshire Hill.
31.—Willow Road.

Plate 11.

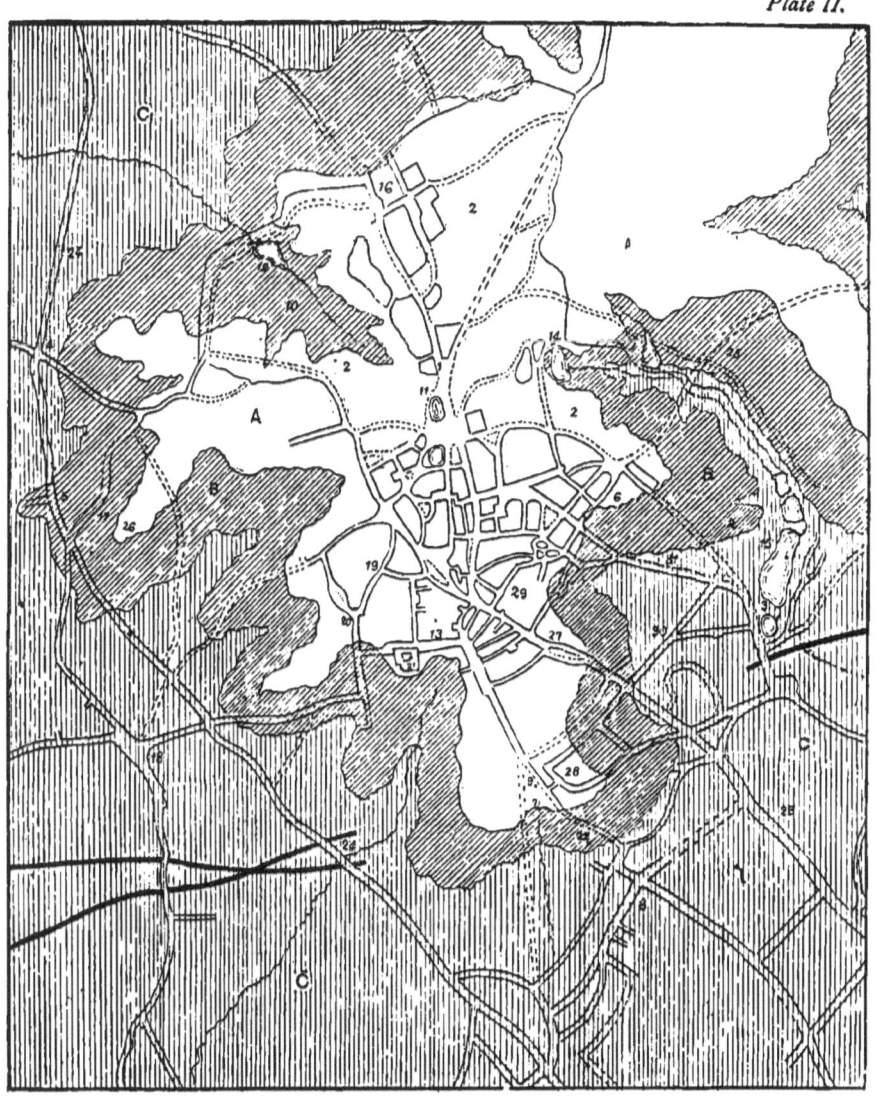

MAP TO ILLUSTRATE THE GEOLOGY OF HAMPSTEAD.

CHAPTER II.

THE hill of Hampstead and its structure, with its consequent surface features and characteristics, are so interesting and instructive that the locality has become, as it were, classic ground for London geologists, who have often assembled on the Heath to listen to a dissertation on the ground beneath them. Dr. Wetherell, Professor Prestwich, Professor Morris, Mr. Whitaker, and Mr. Caleb Evans, and other well-known geologists, have made Hampstead the theme of numerous papers and lectures. The last-named gentleman, the late Mr. Caleb Evans, F.G.S., a resident of Hampstead, from his attentive observation of every instructive section displayed in the district through many years, was the acknowledged master of the subject. By him and by the late Professor Morris, Hampstead Heath was made enchanted ground, for both those genial teachers' greatest pleasure was in imparting to youthful students, on the ground itself, the knowledge of it they so richly possessed. So interesting and instructive, indeed, is Hampstead Hill, that few places in England are more frequently referred to in illustration of physiographical teaching in London schools, for here can be seen an example and an illustration of clay and sand deposits, of hills formed by circumdenudation, of valleys of erosion, of outliers, of the cause of springs, of the formation of marshes, and the sources of rivers, as well as of the dependence of plants

and vegetable growth upon the character of the ground beneath
the surface. How useful it would be if the German custom
of taking the pupils of schools out to see and study instructive
localities were followed in England. Not only would it be
immediately instructive, but it would promote habits of
observation of general educational value, and, at the same
time, greatly interest and delight the pupils, diversify, and
make more attractive their school life, and thereby give them
greater zest and energy for their general work and study.

But, although Hampstead is highly interesting and instructive, the structure of the hill is very simple, since the whole
consists of only two great beds of material called "Formations,"
one lying on the other almost horizontally. The lower one,
which forms the great bulk of the hill, is an immense mass of
clay, and the upper one, which forms the heath-covered summit,
consists of sands ; thus the whole hill is made up of clay and
sand.

The former of these two great beds, the clay, is part of
the main mass of the formation called the "London Clay,"
which extends eastwards from Hungerford, in Berkshire, to
the sea-coast at Harwich, and constitutes by far the greater part
of the extensive area formed by the two counties of Middlesex
and Essex. The other and upper of the two great beds, the
sand, is, on the other hand, a detached and outlying portion
of the formation called the "Bagshot Sands," the main mass
of which, as the name indicates, occupies the district around
Bagshot in Surrey, and forms the extensive barren heaths and
commons about that place, Woking, Aldershot, Frimley, and
Sandhurst, but not extending uninterruptedly further eastwards
than Chertsey and Weybridge.

The London Clay forms the whole of the ground of Hampstead up to about 360 feet above sea-level, and supports the

Bagshot Sands which form the uppermost part of the hill, including the area on which the higher portion of the town of Hampstead stands. All the higher levels of the Heath are consequently on the sand, the sterility of which was the cause of so large a tract of land near the metropolis remaining unenclosed. Hence it is that the existence of the Heath and its appropriation to the people for ever as a public recreation ground, is the direct result of there being here a patch of the Bagshot Sands capping the London Clay forming the mass of the hill.

As the extreme elevation of the hill is 443 feet above Ordnance Datum, the Bagshot Sands have here, consequently, a maximum thickness of about 80 feet. The uppermost beds of the London Clay, to the thickness of about 50 feet, are sandy, becoming more so as the junction with the overlying Bagshot beds is approached. There are therefore three more or less distinct deposits : (1) the typical London Clay forming the base and the great mass of the hill ; (2) the Upper Sandy London Clay above, and (3) the sands of the Bagshot Sands formation capping these and forming the summit of the hill.

This simple structure of Hampstead Hill will be at once seen from the diagram, Plate VII., representing an ideal section through the hill and across the valley of the Thames from the Chalk Country of Bucks and Hertfordshire on the north to the Chalk Downs of Surrey on the south.

CHAPTER III.

OF the material of which the great bulk of Hampstead Hill is composed, it will be necessary to speak somewhat in detail, since, although its general features are commonly known from the many excavations for building and other purposes which daily reveal the stiff tenacious character of this great mass of clay, yet its composition, peculiarities, and especially its extraordinary and most suggestive contents, are not by any means so well known as they ought to be. And there is one simple character of importance that is not at all revealed by ordinary excavations, and that is its colour. This may seem a curious and unwarrantable assertion, but it is nevertheless true, since the colour of the clay exposed in excavations that only penetrate the bed near the surface is not the colour of the mass of the clay. The colour usually seen is, as everybody knows, a yellowish brown, but this is due to a change that has taken place from contiguity to the surface which has allowed the oxygen of air and water to act on the iron matter disseminated through the clay in a manner precisely analogous to the familiar rusting of iron by water or dampness. Thus it is that only the uppermost beds are of the well-known brown colour, while the deeper beds that have been protected from this oxidizing influence are of quite a different hue. The true unaltered colour of this lower clay is a greyish blue, which

Plate III.

"JACK STRAW'S CASTLE" (Summit of Hampstead Hill).

may be seen whenever a deep excavation or well-sinking penetrates to a considerable depth below the surface.

The change of colour produced by the complete oxidation of disseminated ferruginous matter is conspicuously seen when bricks or tiles made of blue clay are burnt, red or reddish bricks or tiles being the result. The iron matter in the unaffected deep clay is, however, an oxide, but only a first oxide, that is, it is a compound of iron with less oxygen than it is capable of combining with, and is hence called a protoxide, but the compound that gives the brown colour is one in which there is a larger proportion of oxygen and it is hence commonly called a per-oxide. It has, moreover, in addition to the oxygen, a certain amount of water combined with it, and so is called a hydrated per-oxide. But when blue-clay bricks are burnt, the iron is not only thoroughly oxidized, but is rendered anhydrous or devoid of water, and this it is that produces the bright red, not brown, colour which is the delight of modern architects and now so much in favour, and perhaps nowhere more so than at Hampstead, where in Fitzjohn's Avenue and in the new buildings of the High Street we may always see a conspicuous illustration of this interesting and beautiful chemical phenomenon.

The substance of the clay, as indeed of all clays, is a compound of a very different character. It is none other than a combination of the substance of sand, or silica, or quartz, the same as rock crystal or "pebbles," with an oxide of that very light, silvery metal we know as aluminium. Thus it is that the basis of the clay is alumina. Now, although aluminium is a light, white metal, alumina is a mineral of the hardest and most intractable character, the next hardest, indeed, to the diamond. In its commoner form it is called corrundum, and its powder emery, but, in its finest and brightest form, in its apotheosis so to speak, it is none other than the ruby and the sapphire. So we find that the substance of these splendid gems of almost adamantine hardness, is the basis of the soft and despised clay which, in wet weather at all events, we so much dislike our feet to come in contact with. The pure silicate of alumina is kaolin, a fine white clay called "china-clay" from being used for the manufacture of porcelain, and this, when mingled with impalpable silicious powder, with such compounds of iron as have been mentioned, and with other substances in a finely divided condition, and permeated throughout with a certain amount of water, constitutes ordinary clay ; and such is therefore the composition of the London Clay of Hampstead.

But besides the substances that make up the body of the clay itself, and only separable by chemical analysis or separately visible by microscopic aid, there are some which form considerable embedded masses, but quite separate and distinct from the clay, and some in such quantities as to be of economic value. One of these distinct substances appears in the form of large, roughly circular masses, sometimes as much as three feet in diameter, and of considerable thickness in the centre, but gradually diminishing towards the circumference, thus forming bi-convex lens-like bodies. These are called " septaria " or " cement stones," the latter name being given to them because they are employed for the manufacture of the so-called " Roman cement," which was so much used in pre-Ruskin days for stuccoing the outsides of brick-built houses, their value for this purpose being due to the material of which they are composed being an admixture of calcareous and clayey matter. They are formed by the natural concentration and aggregation of the calcareous matter contained in the surrounding clay ; and, on account of internal shrinkage, radiating fissures or cracks from the centre with transverse minor cracks are produced, and these become filled up by calc-spar deposited by its crystallization from solution in water which has reached them by percolation from the exterior. Septaria, therefore, on being cut through, present a number of divisions bounded by *septa* (hence the name) or partitions of calc-spar, and the hard substance forming the mass between taking a dull polish, some are available for small ornamental table-tops.

Beautiful crystals of Selenite or crystallized sulphate of lime are also very abundant in the more exposed portions of the clay, where this substance is separated from the clay, or rather produced, by a natural chemical process. The crystals are often almost transparent, and sometimes perfectly so, frequently double, and quite regular in form, and are usually in rhomboidal prisms with bevelled sides. Selenite is the crystalline form of gypsum and alabaster, and is much softer than calc-spar, which is the corresponding crystalline form of carbonate of lime.

Iron Pyrites, a compound of sulphur and iron, although not an abundant mineral in the clay of Hampstead, is so in the London Clay of other areas, especially the Isle of Sheppey, where, from the rapid destruction of the cliffs by the sea, it is collected in sufficiently large quantities from the shore to supply a manufactory of copperas. It frequently encrusts fossils, but often quickly decomposes on exposure, and so destroys the specimens. This mineral is very abundant in Nature, and sometimes occurs in beautiful crystals, and so gold-like that they are often eagerly secured under the belief that they contain the precious metal. Iron Pyrites is, however, of little commercial value, and is not used as an ore of iron.

During the progress of the Archway excavation at Highgate, a curious soft resinous mineral was found in the London Clay, and called " Highgate Resin," and from its resemblance to the gum copal, " Copaline."

CHAPTER IV.

THE London Clay, as a rule, is too stiff for ordinary brick-making operations, but where it has been exposed to weathering action for a long time it may be used for brick and especially for tile-making. The Upper Sandy London Clay is, however, extensively employed, as at the large brickyard below Caen Wood. The ordinary clay is used at West Hampstead and at Willesden Green, and a brickyard some years since was worked at Child's Hill. Brick Earth of the Thames Valley, a much newer deposit than the London Clay, is, however, the usual material employed at the brick-works near London.

The primary origin of clay is to be found in the decomposition of the felspar composing in part granitic and other felspathic rocks, by the action of the carbonic acid gas of the atmosphere, but a knowledge of the conditions of the accumulation of any particular bed of clay forming, in whole or in part, what is known as a "geological formation" can only be approached by attention being given to the organic remains found embedded in it, and by an acquaintance with its geographical extension and relative position to other formations. Though the London Clay bears that name on account of its constituting so much of the district around the metropolis, it has a much greater extension than the neighbourhood of London or even of the Thames Valley.

Plate IV.

OLD HOUSES IN CHURCH ROW.

The London Clay is an important member of the Tertiary or newest of the three great divisions of the Sedimentary rocks. In England the Tertiary rocks (clays and sands are "rocks" in geology) occupy now two distinct areas, separated by a considerable distance from each other. These areas are called " Basins " from the beds having an inclination from both sides of the area towards the interior, and thus we have the " London Basin " and the " Hampshire Basin." The latter includes the southern portion of Hampshire and the northern portion of the Isle of Wight, and is separated by the Chalk area between Winchester and Basingstoke from the London Basin which extends eastward from the foot of the Marlborough Downs, both south and north of the Thames, to the sea, and, broadening eastwardly, it forms a more or less triangular area between the Chalk of Hampshire, Surrey, and Kent on the south, and the Chalk of Bucks, Herts, and Suffolk on the north.

The whole of the Tertiaries are divided into the Eocene, or lowest and oldest, the Miocene, the Pliocene, and the Post-Pliocene, the uppermost and newest, and the London Clay is a member of the lower division of the Eocene group of strata. The lowermost formation of the Eocene in England is called the Thanet Sands, and reposes on the Chalk the uppermost of the Secondaries. The Thanet Sands are followed by the Woolwich and Reading Series, and in Kent by the Oldhaven Beds. Overlying these formations, which are collectively named the Lower London Tertiaries, is the London Clay, a much more considerable deposit than either of the others. This vast accumulation of argillaceous matter extends, as before stated, from Hungerford in Berkshire to the German Ocean, its most eastern point being at Harwich, its greatest southern extension is near Canterbury, and its most northern at Sudbury on the northern border of Essex.

It occupies parts of Berkshire, Hampshire, Surrey, and Kent, forms almost the whole of Middlesex, and by far the greater part of Essex.

Thus we see that the clay of Hampstead Hill is by no means an isolated mass, but that it is merely an elevated portion standing above the general level of the present surface of a very extensive formation, the whole of which must therefore have been produced by deposition in one sea and during one epoch of time.

VOLUTA.

NAUTILUS.
(Greatly reduced.

CASSIDARIA.

ROSTALLARIA.

PECTUNCULUS

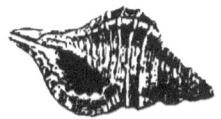

TRITON.

HAMPSTEAD SHELLS.
(*Fossils from the London Clay.*)

All the above figures except that of the Nautilus are drawn from specimens taken from the clay of Hampstead by the Author.

3

CHAPTER V.

ABOUT fifty years ago a few young men, ardent naturalists and at the same time sociable and friendly, as ought always to be the case with lovers of Nature, formed the "London Clay Club," and devoted themselves not merely to convivial suppers but also to the study of the great bed of clay on which London stands and especially to that of its fossils. The late Professor Morris, the first occupant of the Goldsmid Chair of Geology at University College, was one of the little band, so was Dr. Bowerbank of fossil-sponge fame and founder of the Palœontographical Society, and so was Dr. Wetherell of Highgate. To the last-named gentleman all geologists and naturalists are greatly indebted for what we know of the "life of the period," as geologists say, that is, of the creatures that lived at the time that the London Clay was being deposited at the bottom of the sea, for during the time Dr. Wetherell lived at Highgate the Archway Road was formed, and, this cutting through the hill at the level of the most fossiliferous zone of the London Clay, was the cause of a very large number of fossils being disinterred.

With assiduous care the Highgate member of the London Clay Club collected and preserved every fossil he could obtain from the Archway cutting, not only by his own hands, but also by stimulating the friendly aid of the workmen, who were delighted to keep for "the doctor" every shell and every fragment of organism that pickaxe and spade revealed. Thus have been preserved for the instruction of future generations (the collection is now safe practically for all time in the British Museum) the remains of the animals that lived where London now stands, long before those Alpine peaks we call "the everlasting hills" were in existence.

These animals were all marine; they all lived in a salt-water sea such as washes the London Clay shores on the coast of Essex to-day. And that sea must have continued receiving muddy sediment during an immense period of time, for the accumulated mass was 500 feet thick.

The present thickness of the London Clay varies very much, local thinning being due to eroding or denuding agencies that have taken away to the sea by means of rain, streams, and rivers, from areas which are now comparatively low-lying, the greater portion of the original thickness. Besides, however, local thinnings, there is a general diminution of the thickness of the deposit towards the west. The greatest thickness occurs in the Isle of Sheppy, where it is found to be about 470 feet. At Hampstead, where the London Clay is overlain by the next succeeding formation, the Bagshot Sands, there is presumably the full original thickness of the deposit in the Middlesex area, and it is here about 400 feet thick.

The data given by the details of the sections exposed by wells that have been sunk in various parts of the district show clearly that the thickness depends upon the elevation above sealevel, and that therefore the thinning process has taken place on the top, and, consequently, has been due to denudation, or the wearing away of the surface by weathering action, which at a few points has spared portions of the next newer deposit, and so the original thickness of the London Clay at those places is shown. Thus at Whitaker's Brewery, Camden Town, the London Clay is 140 feet thick ; at Camden Town Station (L. & N. W. R.), 144 feet ; at the Zoological Gardens, Regent's Park, 155 feet ; at Cricklewood, 212 feet ; and at the old well at the bottom of the Lower Heath, 289 feet.

At Hampstead Heath, at Harrow, and at High Beach in Essex, remnants of the Bagshot Sands form the summit of the land, and hence at these points the full thickness of the London Clay in those districts remains. At the Crystal Palace, Sydenham, the whole of the once overlying Bagshot Sands has been taken off, but a portion of the uppermost bed of the London Clay, the Upper Sandy London Clay, caps the hill, so that had Sydenham Hill not been worn down quite so much it would have had at the summit a patch of Bagshot Sands like Hampstead Hill.

Since the London Clay forms a member of the Tertiaries of the Hampstead Basin, it may be safely concluded that it has at one time extended over the Chalk area lying between the two Tertiary Basins. There has therefore been a continuous bed of the London Clay from Harwich to the Isle of Wight, where portions may be seen in White Cliff and Alum Bays. But that was not its extreme extent by any means, for it is represented on the other side of the Channel, in France, near Dieppe, and also in the neighbourhood of Dunkirk, while some beds in the Belgian area appear to be of the same age, and were doubtless deposited in the same sea.

What changes in the geography of this part of the European area are told to us by this superficial glance at the extension of the London Clay ! for it tells of continuous continental land without the English Channel, without the straits of Dover, without the white cliffs of Albion, and without the Solent and Spithead to insulate the Isle of Wight. But it tells much more than this and of an earlier epoch. It tells of the time when the whole of the area it now occupies was sea, and from its utmost extension we learn the minimum extent of the sea in which it was deposited.

And still more even than this may be learned, from the fossils which are embedded in it, of geographical changes. For, since they show the kind and character of the organic life of the London Clay sea, so also do they indicate clearly to us the general climatal conditions of the area of the sea ; or, in other words, they tell us whether the sea was a warm one or a cold one.

The London Clay furnishes an abundant and interesting fossil Fauna, and a less abundant but still interesting fossil Flora. The fossil Flora, or assemblage of plant remains, is almost exclusively confined to the Isle of Sheppey area, where large quantities of fruits have been found fossil. The fruits have evidently been similar to those now growing on the banks of tropical rivers, and which at certain seasons almost cover the surface of their waters. Many species have been described by Dr. Bowerbank from the Sheppey London Clay, but the genus best represented greatly resembles the *Nipa*, so abundant on the banks of the Ganges, and is therefore called *Nipadites*. The inference from the abundance of these remains of land plants in the Isle of Sheppey, and their limitation to that area, is that in that neighbourhood debouched a river, which, after running through lands covered with palms and other tropical vegetation, brought the fruits which had fallen on its surface to the sea, to the bottom of which they in due course sank and so were embedded in the accumulating sediments.

The fossil Fauna of the London Clay, though abundant, is subject to remarkable restrictions, both as to area and vertical position. Thus the higher Classes, those having backbones, or the Vertebrata, the reptiles and fishes in fact, are almost exclusively confined to the Kentish area, while the distribution of the lower

Classes, those without backbones, the Invertebrata, or what are commonly, though very incorrectly, called "shell-fish," to which nearly all the fossils of the Hampstead area belong, is thus described by Professor Prestwich : " It would appear that although a great proportion of the fossils range at intervals vertically throughout the London Clay, yet their development is very different in different zones, being abundant in some, and scarce in others, whilst each zone is further marked by a few characteristic species ; thus forming distinct, although nearly related groups."

The Vertebrata consist chiefly of reptiles and fishes. So numerous are the remains of turtles in the London Clay of Sheppey that no less than seven species have been described, although only two species are now known to be living in all the seas of the globe. In the class of fishes, eighty-eight species are recorded from the Sheppey area and only six or seven from the Hampstead district. Many of these were quite similar in character to fishes now living though not the same in specific detail.

Of all the localities in the Middlesex area, Highgate, as has been intimated, has proved by far the most prolific of organic remains. The very large number of 180 named and described species were obtained from the Archway excavation by Dr. Wetherell. These were disentombed from a bed about 130 feet below the level of the summit of the hill, doubtless the same bed which has proved to be prolific of fossils at Hampstead but probably lower than the Sheppey zone. This bed was cut into in 1871, when a sewer was being constructed along a portion of the Finchley Road at Child's Hill. Mr. Caleb Evans obtained sixty species from this excavation. He described the beds as consisting of a yellowish clayey sand, to the depth of 12 feet, with few fossils, but passing down first into a dark-grey clayey sand and then into a sandy clay, at the base of which was ordinary stiff clay. In the sandy clay fossils were in great abundance, but of these a few species very largely predominated. By far the most abundant were a *Voluta* and a small bivalve shell, a *Pectunculus*. I myself, in a very short time—less than half an hour—obtained from this excavation over a hundred of these shells. Though not found at this spot the beautiful Nautilus shell is characteristic of this zone, and was conspicuous at High-gate. From this it will be seen that this fossiliferous bed is at about 130 feet below the summit of the Heath, or about 300 feet above sea-level.

An excavation in high ground near Child's Hill, in 1866, was also described by Mr. Evans. Here the base beds of the Bagshot Sands gradually became more clayey and contained a great amount of water. Lower down these beds passed into the sandy clay containing again the Voluta and the Pectunculus in great numbers. This sandy clay extended along Platt's Lane and the Finchley Road nearly to New West End. Between Child's Hill and North End the sewers traversed the water-bearing stratum, but at one spot, in the swampy ground by the " Leg of Mutton Pond," the fossiliferous bed was reached. Mr. Evans writes : " A similar succession of strata was seen in 1862 in drainage works in Frognal Lane. The upper part of this exposure showed the yellow Bagshot Sand at Frognal House. Lower down the lane, near the entrance to Oak Hill Park, the dark-grey sand was seen, and at the corner of the lane leading to the Parish Church and near the Priory, the sandy clay." Let us hope that there are many residents in Hampstead who will emulate the habit of observation and the single-minded desire to record facts of Nature which Mr. Evans so markedly possessed.

Plate VI.

"THE SPANIARDS," HAMPSTEAD HEATH.

CHAPTER VI.

OF our children who pick up shells on the sea-shore and bring them home to Hampstead as trophies of their, to them, great but delightful travels and adventures, how few there are who know that sea shells are in the clay on which their homes are built. But the Hampstead shells are like those they only see in the sea-side shop, and cannot find with all their industrious search on the beach below, for they are similar to the beautiful West Indian shells so temptingly displayed for sale, and thus these shells teach the interesting fact that warm climatal conditions prevailed during the formation of the clay of Hampstead, for living volutes and nautili we know are only to be found in tropical or sub-tropical seas. This conclusion is confirmed by the remarkable fossils of the Isle of Sheppey. The many remains of turtles there found, together with the abundance of fossil tropical-like fruits have led some to the conclusion that torrid conditions obtained, but the small size of the volutes and other warm sea shells seem to indicate rather a moderately warm or sub-tropical sea, but certainly one much warmer than that which now washes the shores of the British Islands. Although not a very deep sea, there is reason for supposing it to have had a depth that would allow of a considerable variation of temperature, and have much warmer water where shallow near the shore than at its greater depths, by which both sub-tropical and temperate marine creatures might find waters of the temperature they required.

The warmer climate of this area at the period of the formation of the London Clay has formed the subject of much discussion, and various ways of accounting for it have been suggested. Some suppose a change in the position of the earth's axis, some a greater heat of the sun ; but it should be remembered that changes of geographical outlines of land and sea produced by those slow and long-continued upward and downward movements of the earth's surface, which, in some region or another, have always been going on, will account for much climatal alteration. By supposing a sea closed to all northerly currents of cold water, and open to the south-west by which it would receive the full benefit of those warm flows from the mid-Atlantic which even now raise the temperature of the water on our southern and western coasts, we may easily account for a very considerable increase of temperature at a particular epoch.

Plate VII.

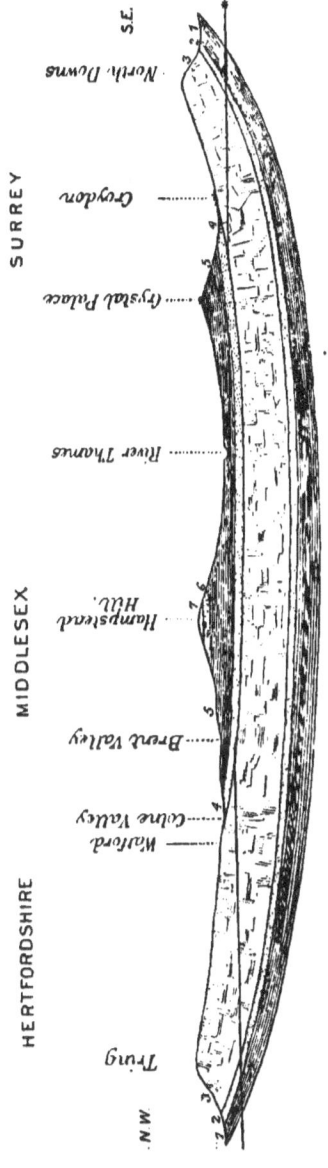

HERTFORDSHIRE MIDDLESEX SURREY

N.W. S.E.

Tring Watford Brent Valley Hampstead Hill River Thames Crystal Palace Croydon North Downs

Colne Valley

3. The Chalk, Chalk Marl (Lower Chalk, and Upper Chalk with Flints).
2. The Upper Greensand (Sands with green grains and calcareous stone).
1. The Gault (Blue Clay).
* Sea level.

7. The Bagshot (Sands Hampstead Outlier).
6. The London Clay (Upper Sandy Bed).
5. The London Clay (Main Bed).
4. The Lower London Tertiaries (Thanet Sands and Woolwich Beds).

SECTION FROM BUCKINGHAMSHIRE TO SURREY.

(Showing the structure and sub-structure of Hampstead Hill and the Thames Valley.)

CHAPTER VII.

It is now time, however, to say something of the sands which give so marked a character to the summit of the hill. They are so well and conspicuously displayed, and therefore so well known, that a description of their appearance is scarcely necessary. Their prevailing colour is yellow of various shades, but, like the clay, they undergo a change in this respect with exposure to weather—not, however, by increased oxidation, but by having their colouring matter removed. The yellow sand is coloured, as is, indeed, all brown and red sand, by each grain being coated over by a thin film of iron-oxide, the grain itself being a particle of white quartz or colourless rock-crystal; in each case, however, silica—a grain, in fact, which is a fragment of some old granitic or quartz-bearing rock or quartz rock itself, some quartz vein, or some previously existing bed of sand or sand-stone. When, therefore, this thin film or coat of colouring matter is worn off, or taken off by exposure to the weather, the sand becomes white or whitish. In some places there is sufficient iron-oxide to cement the sand together into a ferruginous concretion, producing irregular-shaped masses of varying hardness. Occasionally, too, some thin clayey seams are to be found interstratified with the sands. The grains themselves are more or less rounded or sub-angular, as it is termed, giving evidence of some amount of wearing action on a sea-shore.

It is stated that a fragment of a fossil shell was found in a ferruginous concretion in the sands at Hampstead, but with this exception there is no record of organic remains having been found in these beds, and this is quite in accordance with the unfossiliferous character of the sands of the same age in other areas. By remains of organisms, therefore, the sands do not afford evidence of the climatal conditions under which they were deposited, though they themselves tell us something of the geographical changes that were occurring in this part of the world at the time of their deposition and immediately preceding it, but the fossils and position of clay and limestone beds of similar age in England and abroad reveal to us a wonderful story.

Grains of sand are much larger and heavier individually than the particles forming clay, and consequently sink to the bottom of water in which both may be suspended earlier, and therefore in the sea nearer to the shore than the very minute clay particles. Hence these sands were deposited in shallower water than the clay below, and, as they are in the same area, it is evident that the sea bottom must have risen between the clay period and the sand period, and that this elevation of the sea bottom and consequent shallowing of the sea was gradual is told by the upper beds of clay becoming more and more sandy until the sands, pure and simple, are reached. Thus we read and learn from the simple fact of the sands of Hampstead being above the clay of Hampstead, the geographical changes that took place in this area long long before history began to be written.

The sands on the Heath have been excavated for a long period for economical purposes, and consequently the old natural level has not been preserved. An enormous amount of sand has thus been artificially removed from the north-west side, giving a much lower level than that of the summit-road, and leaving some of the older trees standing on isolated mounds with the roots in many cases exposed. The picturesque clumps of Scotch firs which have attained large dimensions, attest the suitability of these trees to the ground on which they grow so well, while the oak and elm are to be looked for on the lower levels where clay and sandy clay form the subsoil.

As has been previously stated, the capping of sands on Hampstead Heath is an outlying patch of the Bagshot Sands, quite cut off from the main mass of the formation which occupies an extensive area in Surrey, Berks, and Hants. The Bagshot Sands are everywhere very sterile, and are chiefly covered with heaths and commons and pine-woods. They give to us those extensive wastes round about the junction of the three counties just named that have been found so useful for military and other purposes requiring large areas of ground of little or no agricultural value.

The "Hampstead Outlier," as it is called, is one of the best examples of an outlier since it is so distinct from the surrounding beds, so limited in area and so well exposed. Its presence, besides marking the top of the maximum thickness in this area of the underlying London Clay, indicates the original extension of the great sheet of the Bagshot Sands to the north side of what is now the Thames Valley, and the amount of denudation that has taken place between this point and the Surrey beds, by which the present levels and form of the ground of the London area have been produced.

From this it is seen that in the middle of the Thames Valley about 400 feet thickness of solid material has been removed to form the present bed of the river. The amount of material removed lessens as the northern or southern boundaries of the Thames Valley are approached, and at Hampstead, Highgate,

4

and Harrow, we find the relative level of the previously continuous and far-extending sheet of super-imposed and overlying Bagshot Sands.

Hampstead is not, however, a portion of the northern
boundary of the Thames Valley, but a hill which, with the
adjoining one of Highgate, forms a hill-mass standing within
the great Thames Valley some distance to the south of the
bounding ridge, and, having evidently been the result of denudation on all sides, forms a good example of a hilly mass left
standing above the surrounding country by circumdenudation.

The Bagshot Sands generally constitute the middle division
of the English Eocene Tertiaries, which is represented by most
important massive and wide-spreading rocks on the continent
of Europe and in Asia and Africa, for to the Middle Eocenes
belong the well-known Paris building-stone, the *Calcaire gros-
sier*, extensive beds in India and the Nummutitic Limestone of
Egypt used by the builders of the Pyramids. And as rocks of
later age form portions of the Alps as high as 5,000 and
6,000 feet above the level of the sea, as well as large portions
of the Himalayas, the highly-interesting and important fact is
ascertained that the sands of Hampstead Heath and, indeed,
all the materials of Hampstead Hill were deposited before the
most prominent features of the present surface of the Old World
came into existence.

On the surface of the highest part of the Heath, near the
Spaniards, will be found some rounded pebbles, thought to be
remnants of gravels similar to those which lie on the hill-tops
on the north of Middlesex, and named by Mr. Whitaker "Pebble
Gravel." They are doubtfully attributed to glacial action, and
deemed to be the oldest of the Glacial Deposits. One or two
patches on the northern side of the hill have been referred to
the Glacial Period, but, with these possible exceptions, the Glacial

sands and clays seem to have been denuded from the Hampstead area. Undoubted Glacial beds occupy extensive areas in the neighbourhood of Finchley and Hendon, and at Muswell Hill, and on the west at Ealing.

CHAPTER VIII.

SUBSEQUENT to the Glacial Period, the extensive denudation occurred that completed the excavation of the great Thames Valley, as well as that of the subsidiary and lateral valleys of the Brent, the Yedding, and the Colne, in the middle and west of Middlesex, and of the Lea on the eastern side of the county. The land in this district was thus left sculptured as we now see it, with Hampstead, Highgate, and Harrow Hills standing in the midst, and preserving remnants of the great sheet of Bagshot Sands.

The sculpturing produced by water action is beautifully seen around the Heath. From the summit, valleys radiate to the north-west, to the south-west, to the south-east, and to the east. Each of these valleys gives rise to springs in its upper part by the outflow of water from the base of the sands, consequent upon the stoppage of vertical percolation by the impervious clay beds below. From these outflows of the water of the summit-sands are produced sources of a feeder of the Brent, of the Bayswater and Westbourne, and of the Fleet River. Of these streams the affluent of the Brent is the only one now flowing along its original natural channel, which may be seen below the Leg of Mutton Pond, forming a very beautiful example of a "valley of erosion."

Just below the junction of the Bagshot Sands and the London Clay, and on the area occupied by the Upper Sandy Clay, will be found in the valley on the West Heath commencing below the flag-staff, near Jack Straw's Castle, and opening out to the north-

Plate VIII.

VALE OF HEALTH (Highgate in the distance).

west some marshy ground. This has recently been partly drained, and so deprived, to some extent, of its natural character, to the great regret of many naturalists, but still, especially in the winter, a true river source is well exemplified. The drainage of the uppermost part of this valley is collected in the Leg of Mutton Pond, and afterwards passes on through the miniature valley of erosion before-mentioned to feed the Brent. The old Bayswater and Westbourne stream had its source in the south-west or Oak Hill and Frognal valleys, and the sources of the Fleet on the south-east now supply the Vale of Health Pond and the ponds on the Lower Heath.

Besides the streams above-named, the sands of the Heath supply by their drainage the water for several well-known springs, which issue on the town side of the hill and mark the uppermost level of the clay. The Conduit Spring near to Rosslyn Bank was long known as a source of good water, but is now injured by the shaft of the railway tunnel. The chalybeate spring near Well Walk has been famous since the beginning of the last century. The chalybeate character of the water is due to the iron oxide of the sands through which the water percolates. It has now but a very small flow, yet the chalybeate impregnation of the water is still very decided. The occurrence of springs or outflows of water round the hill will therefore serve, where no exposure of the beds are to be seen, to determine the approximate position of the junction line between the pervious Bagshot Sands and the underlying London Clay. It follows also that dampness of foundations at a high level, a cause of surprise to many, will be easily accounted for by the geological conditions and structure of the hill. This important line of junction may, however, be actually seen at various points where exposures of the lower strata of the sands occur, notably at North End, at Oak Hill Park, and at the brickyard between Hampstead and Highgate.

CHAPTER IX.

REGARDING Hampstead Hill as standing on a base 100 feet above Ordnance Datum, its substructure or foundations, so to speak, for 1,000 feet in depth is a composite one, as is shown by the vertical section (Plate IX.). There is first the lower portion of the already-described London Clay with its basement bed, 400 feet, then the Woolwich and Reading Beds below of 60 feet, with the underlying Thanet Sands of 30 feet thickness. Below all these beds lies the great Chalk formation, having here a total thickness of about 650 feet, but divisible into the Upper Chalk with flints, 250 feet; the Lower Chalk without flints, 340 feet; and the Chalk Marl, 60 feet. At the base of the Chalk series there is a comparatively thin bed of 20 feet, called the Upper Greensand, and this rests upon the Gault Clay, which makes up the remainder of the 1,000 feet of strata underlying the base of Hampstead Hill.

The Woolwich and Reading Beds are very variable, but yet display in the western, or Reading, area a decided difference of character from that of the beds in the eastern, or Woolwich, area. This difference is sufficiently distinct and important, since it indicates a geographical difference of conditions of deposition, to justify the use of the two names, Woolwich and Reading, for the designation of the formation as a whole. The fossils of this

Plate IX.

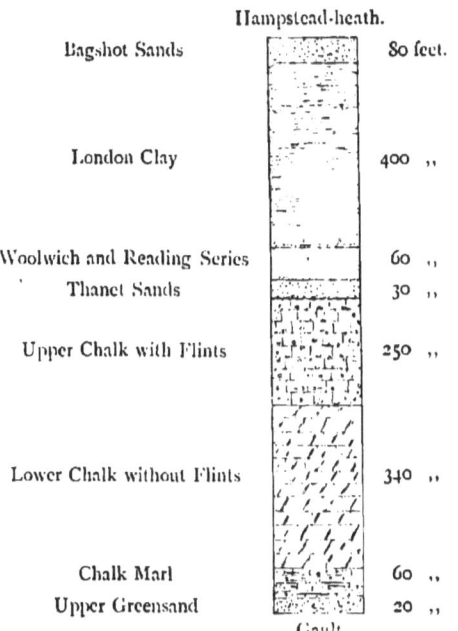

Hampstead-heath.

Bagshot Sands	80 feet.
London Clay	400 ,,
Woolwich and Reading Series	60 ,,
Thanet Sands	30 ,,
Upper Chalk with Flints	250 ,,
Lower Chalk without Flints	340 ,,
Chalk Marl	60 ,,
Upper Greensand	20 ,,

Gault.

VERTICAL SECTION UNDER HAMPSTEAD HEATH.

(Showing the formations from the Surface to the Gault.)

series clearly lead to the conclusion that the material forming the beds was deposited under both marine and estuarine conditions, the latter prevailing in the eastern area, where, at Charlton, near Woolwich, river and estuarine shells may be found in great abundance; so abundant, indeed, are they, that some beds are almost entirely composed of them.

Although the Thanet Sands are purely marine in origin, they were deposited in a small sea, and have a very limited extension. Thickest in the east (in the Isle of Thanet—whence their name), they gradually thin out to the westward, and do not seem to extend further in that direction than to the neighbourhood of Ealing. There are not many exposures of these beds on the north of the Thames; but on its southern bank at Charlton, Erith, and Upnor, they are well seen showing, by artificial sections—the sands being extensively worked for iron-casting purposes— exposures of buff-coloured loamy sands 70 or 80 feet thick.

The very important Chalk formation, which has a maximum aggregate thickness of upwards of 1,000 feet, and an extension in England from Dorsetshire to the Forelands and from the South Coast to Flamborough Head, is of such a remarkable character that much could be said about its origin, its composition, and its structure. It may, however, suffice for these pages to state that it is a great accumulation of calcarious matter formed by minute animals that lived in a deep and open sea, such as the present Atlantic Ocean, in which it has been ascertained that a similar deposit is now being formed.

This Chalk underlying the Tertiary beds below Hampstead Hill, as will be seen from the section (Plate IX.), is not only con- tinuous with the chalk of Watford and Hertfordshire on the north, but also with the chalk of Croydon and Surrey on the south. Dipping to the south, it extends quite under the valley of the Thames, and then, gently rising, comes to the surface at

Croydon, thus forming a basin in which repose the later-formed Lower London Tertiaries and the still later London Clay.

Underlying the whole of the Chalk, the Upper Greensand and the Gault extend quite comformably, as geologists say ; that is, they have the same angle and direction of dip, and are consequently parallel with the Chalk above. The former of the two, the Upper Greensand, though in the main sandy, has some valuable beds of stone, partly calcareous, called " Fire-stones." The green grains giving the name " Greensand " consist of a silicate of iron.

A more important formation is the Gault, which has a thickness in some places of 200 feet, and, being a fine, blue clay, is largely worked, where it is at the surface, for the manufacture of tiles and red bricks.

It is the basin-like extension of the porous, and therefore water-bearing and water-yielding, Chalk overlying a great bed of impervious clay (the Gault), and covered by a great bed of impervious clay (the London Clay), that constitutes this formation the great reservoir of water available for the supply of the inhabitants of the Thames Valley whenever wanted ; and it is this geological structure of the valley of the Thames that gives the conditions necessary for the formation of those " artesian wells " which at present furnish so easily and so continuously an abundant supply of the purest water to our great breweries and other large manufactories. The water of the concave porous chalk strata lying between the impervious beds of clay rises freely in borings at places lower than the level of the super-saturation of the chalk where that rock forms the surface, and where it receives the water from the rainfall of Hertfordshire, Kent, and Surrey.

WELL SECTIONS IN THE NEIGHBOURHOOD OF HAMP-STEAD.[1]

THE following measured sections obtained by well-sinking will show the ascertained thickness of the beds below eight points in the neighbourhood of Hampstead. By far the most important is the Kentish Town section, obtained by a boring for water for the New River Company, which gives the thickness of the beds to a depth of 1,302 feet. A summary of the details of this section, taken from Professor Prestwich's paper in the *Quarterly Journal of the Geological Society*, vol. xii., gives the following measure ments as arranged by Mr. Whitaker :—

KENTISH TOWN WELL.

Surface of the ground 174 feet above Thames High Water Mark.
Shaft 539 feet, the rest bored.

	Ft.	In.
London Clay . . .	236	0
Woolwich and Reading Beds	61	6
Thanet Sands . . .	27	0
Chalk with Flints (Upper Chalk) .	244	6
Chalk without Flints (Lower Chalk)	341	0
Chalk Marl . . .	59	3
Upper Greensand . . .	13	9
Gault	130	6
Sands, Sandstone, Clayey Sands, Clays, and Conglo- merates of doubtful age, prevailing colour red .	188	6
Lower Greensand (?)		
Total . .	1,302	0

[1] From the " Geology of the London Basin," by William Whitaker, B.A., F.G.S.

CAMDEN STATION, London and North-Western Railway.   About 100 feet above
    Ordnance Datum   Sunk 180 feet, the rest bored.   Water rose to a height
    of 150 feet below the ground.

	Ft.	In.
Made Ground	18	0
London Clay	144	0
Woolwich and Reading Beds	64	0
Thanet Sands	8	0
To Chalk	234	0
In Chalk	166	0
Total	400	0

CAMDEN TOWN, Whitaker's Brewery.

	Ft.	In.
London Clay	140	0
Lower London Tertiaries	70	0
To Chalk	210	0

CRICKLEWOOD, near Hampstead.   157 feet above Trinity High Water Mark
    Water level about 110 feet down.

	Ft.	In.
Loam and Gravel	13	0
London Clay	212	0
Plastic Clay	45	0
Thanet Sands	21	0
To Chalk	291	0

HAMPSTEAD, LOWER HEATH.   About 230 feet above Ordnance Datum.   Sunk
    320 feet, the rest bored.

London Clay, 289 feet :	Ft.	In.
Brown Clay	30	0
Blue and dark brown Clay with septaria	170	0
Clay sandy and with vegetable remains	84	0
Basement Bed, Rock with green grains and many fossils	5	0

Lower London Tertiaries, 89 feet :	Ft.	In.
Plastic Clay . .	40	0
Sand . . .	49	0
Thin bed of flints.		
To Chalk .	378	0
Hard Chalk . . .	40	0
Soft Chalk, with water .	4	0
Hard Chalk, no water .	28	0
Total	450	0

HAVERSTOCK HILL, Orphan Working School. About 170 feet above Ordnance Datum. Shaft 230 feet, the rest bored. Water rose to 196 feet below the surface.

	Ft.	In.
London Clay . . .	223	0
Woolwich and Reading Beds	61	0
Thanet Sands	28	0
To Chalk	312	0

KILBURN BREWERY. Water rose to within 150 feet of the ground.

	Ft.	In.
Blue London Clay . .	235	0
Mottled clay, pebbles, and sand	45	0
To Sand Spring	280	0

REGENT'S PARK, Zoological Gardens. ? 128 feet above Trinity High Water Mark.

	Ft.	In.
London Clay . . .	155	0
Reading and Thanet Beds	69	0
To Chalk	224	0

LISTS OF FOSSILS FROM THE LONDON CLAY OF THE PARISH OF HAMPSTEAD.[1]

The following Lists of Fossils (except that of the Child's Hill species) have been compiled from the "Geology of the London Basin," by Mr. Whitaker; the papers in the "Transactions" and " Quarterly Journal of the Geological Society," by Professor Prestwich; the "Monographs of the Palæontographical Society," by Professor Bell, Charles Darwin, F. E. Edwards, Professor Rupert Jones, and Searles V. Wood.

I have also taken advantage of the recently completed splendid arrangement of Type Collections of Mr. Wetherell and Mr. Edwards in the British Museum, facilities for the examination of which have been very kindly afforded me by the chiefs of the Geological Department, Dr. Woodward and Mr. Etheridge. My especial thanks are due to these gentlemen for giving me access to the MS. records of the British Museum not hitherto published, from which many of Mr. Edwards's species have been obtained.

The Child's Hill list is from an examination of the collection of the late Mr. Caleb Evans, together with MS. records which have been most kindly placed at my disposal by his brother, Mr. John Evans. These records have been of service also in the completion of the Haverstock Hill list.

The nomenclature of the Foraminifera is the most recent and given on the authority of Messrs. Sherborn and Chapman, " Journal Roy. Micro. Soc.," series 2, vol. vi.

---

[1] From "The Geology of the Parish of Hampstead," by J. Logan Lobley, F.G.S., &c., in the "Transactions of the Middlesex Natural History Society " for 1886–7.

UPPER SANDY LONDON CLAY.

FOSSILS from CHILD'S HILL. (With the exception of Ophiura Wetherelli, from excavation for sewer in Finchley Road near Burgess Hill.)

Cephalopoda.
Nautilus sp.

Gasteropoda.
Actæon simulatus, Sow.
Buccinum labiatum, Sow.
Bulla constricta.
Calyptræa trochiformis, Lam.
Cancellaria læviuscula, Sow.
Cassidaria nodosa, Brand.
Conus concinuus, Sow.
Dentalium sp.
Fusus asper, Sow. (?)
„   bifasciatus, Sow.
„   complanatus, Sow.
„   læviuscula, Sow.
„   pseudo-porrectus.
„   trilineatus, Sow.
Murex coronatus, Sow.
„   cristatus, Sow.
Natica labellata, Lam.
Pleurotoma abnormis, Edw.
„   denticula, Bast.
„   fasciolata, Edw.
„   parilis, Edw.
„   teretrium, Edw.
„   Wetherelli, Edw.
Pyrula Greenwoodi, Sow.
Ringicula turgida, Sow.

Rostellaria lucida, Sow.
Scalaria sp.
Solarium canaliculatum, Lam.
Triton argutus.
Turritella imbricataria, Lam.
Voluta nodosa, Sow.

Lamellibranchiata.
Astarte rugata, Sow.
Avicula media, Sow.
„   papyracea, Sow.
Cardium nitens, Sow.
Cytherea tenuistriata, Sow.
Modiola depressa, Sow.
„   elegans, Sow.
„   subcarinata, Lam.
Neæra inflata, Sow.
Nucula compressa, Sow.
Panopæa intermedia, Sow.
Pecten corneus, Sow.
Pectunculus decussatus, Sow.
Pinna affinis, Sow.
Solen affinis, Sow.

Brachiopoda.
Lingula tennis, Sow.

Echinodermata.
Ophiura Wetherelli, Forb.
Spatangus sp.

MAIN BED OF THE LONDON CLAY.

FOSSILS from the LOWER EAST HEATH. (Well sinking for Water Company.)

Gasteropoda.
Actæon simulatus, Sow.

Bulimus ellipticus, Sow.
Cancellaria læviuscula, Sow.

5

Cypræa oviformis, Sow.
Dentalium anceps, Sow.
   ,,   nitens, Sow.
Pleurotoma Koninckii, Nyst.
   ,,   Selysii, De Kon.
   ,,   simillima, Edw.
   ,,   terebralis, Lam.
Pyrula Greenwoodi, Sow.
   ,,   nexilis, Lam. (?)
Rostellaria lucida, Sow.
Scalaria reticulata, Sow., not Brand.
Solarium patulum, Lam.
(?) Voluta nodosa, Sow.

Lamellibranchiata.

Anomia scabrosa, Wood.
   ,,   tenuistriata, Desh.
Arca impolita, Sow.
   ,,   nitens, Sow.
Astarte rugata, Sow.
Avicula arcuata, Sow.
   ,,   media, Sow.
   ,,   papyracea, Sow.
Cardium nitens, Sow.
Corbula Regulbiensis, Mor.
Cryptodon angulatum, Sow.
Leda amygdaloides, Sow.
   ,,   partim-striata, Wood.
Modiola depressa, Sow.
   ,,   elegans, Sow.
Nucula Bowerbankii, Sow.
   ,,   compressa, Sow.
   ,,   similis, Sow.
   ,,   Wetherelli, Sow.
Panœpa intermedia, Sow.
Pecten corneus, Sow.
   ,,   duplicatus, Sow.
Pectunculus decussatus, Sow.
Pinna affinis, Sow.
Teredo antenautæ, Sow.

Brachiopoda.

Lingula tenuis, Sow.
Terebratulina striatula, Sow.

Polyzoa.

Cellepora.

Crustacea.

Cythere (?) barbata, Sow.
Pollicipes (?)
Scalpellum quadratum, Darwin.
Xanthopsis Leachii, Desm.

Annelida.

Vermilia crassa, Sow.

Echinodermata.

Cidaris Websteriana, Forb.
Pentacrinus subbasaltiformis, Mill.
   ,,   Sowerbyi, Weth.

Actinozoa.

Dasmia Sowerbyi, M. Edw.
Graphularia Wetherelli, M. Edw.

Foraminifera.

Clavulina (Textularia) communis, d'Orb.
Cristellaria (Nodosarina) cultrata, Montf.
   ,,   ,,   Italica, Defr.
   ,,   ,,   rotulata, Lam.
Dentalina Buchii.
   ,,   communis, d'Orb.
   ,,   elegans, d'Orb.
   ,,   spinulosa, Mont.
Marginulina Wetherelli, Jones.
Miliolina (Quinqueloculina) triangularis, d'Orb.
Nodosaria Badensis, d'Orb.
   ,,   raphanistrum, Linné.
   ,,   raphanus, Linné.
Planorbulina Akneriana, d'Orb.
   ,,   Ungeriana, d'Orb.
   ,,   Haidingeri, d'Orb.
Pulvinulina Boncana, d'Orb.
Trochamm inaincerta, d'Orb.

Fossils from Haverstock Hill. (Excavation for Tunnel of the Midland Railway.)

Cephalopoda.
Nautilus centralis, Sow.
   ,,   (Aturia) zic-zac, Bronn.
Gasteropoda.
Actæon simulatus, Sow.
Bulla sp.
Cancellaria læviuscula, Sow.
Cassidaria carinata, Lam.
Cerithium Charlesworthii, Prest.
   ,,   Londinense, Edw. MS.
Conus concinuus, Sow.
Dentalium acuticosta, Desh.
   ,,   bisiphonatum, Edw.
   ,,   subcaniculatum, Edw.
Eulima angustior, Edw. MS.
Fusus bifasciatus, Sow.
   ,,   coniferus, Sow.
   ,,   interruptus, Sow.
Natica labellata, Lam.
Pisania dubia, Edw. MS.
Pleurotoma dissimilis, Edw.
   ,,   teretrium, Edw., var. tuberculata, Edw.
Pyrula Greenwoodi, Sow.
   ,,   Smithii, Sow
Rissoa ?
Rostellaria lucida, Sow.
Scalaria antiqua, Edw. MS.
   ,,   perforata, Edw. MS.
   ,,   sororcula, Edw. MS.
Solarium canaliculatum, Lam.
Voluta elevata, Sow.
   ,,   nodosa, Sow.
   ,,   Wetherelli, Sow.
Lamellibranchiata.
Anatinella tenerima, Edw. MS.
Anomia scabrosa, S. Wood.

Arca impolita, Sow.
   ,,   nitens, Sow.
Astarte rugata, Sow.
Avicula media, Sow.
   ,,   papyracea, Sow.
Cardium semigranulatum, Sow.
Corbula globosa, Sow.
Cyprina planata, Sow.
Leda amygdaloides, Sow.
   ,,   partim-striata, Wood.
Neæra inflata, Sow.
   ,,   triradiata, Brit. Mus. Cat.
Nucula Bowerbankii, Sow.
   ,   Wetherelli, Sow.
Ostrea gryphovicina, Wood.
Pecten corneus, Sow.
   ,,   duplicatus, Sow.
Pectunculus decussatus, Sow.
Pholadomya margaritacea, Sow.
Pinna affinis, Sow.
Syndosmya splendens, Sow.
Tellina sublavis, Edw.
Teredo antenautæ, Sow.
Brachiopoda.
Terebratulina striatula, Sow.
Crustacea.
Hoploparia sp.
Xanthopsis sp.
Annelida.
Ditrupa sp.
Vermicularia Bognoriensis, Mant.
Echinodermata.
Hemiaster Branderianus, Forb.
Pentacrinus subbasaltiformis, Mill.
Actinozoa.
Dasmia Sowerbyi, M. Edw.
Leptocyathus elegans, M. Edw.

Stephanophyllia discoides, M. Edw.
Turbinolia Prestwichii, M. Edw.
Websteria crisioides, M. Edw.
   Foraminifera.
Dentalina communis, d'Orb var.
   „ spinulosa, Mont.
Marginulina similis, d'Orb.
Miliolina (Triloculina) trigonula, Brit. Mus.

Miliolina (Quinqueloculina) sp.
Nodosaria consobrina, d'Orb.
   „ pauperata, d'Orb.
   „ raphanistrum, Linné.
   „ raphanus v. zippii, Reuss.
   „ soluta, Reuss.
Textularia carinata, d'Orb.

FOSSILS from PRIMROSE HILL. (Excavation for Tunnel of the London and North-Western Railway.)

Aves.
Sternum of a small wader.
   Reptilia.
Trionyx sp.
   Pisces.
Lamna elegans, Agassiz.
Otodus obliquus, Agassiz.
   Cephalopoda.
Nautilus centralis, Sow.
   „ regalis, Sow.
   „ Sowerbyi, Weth.
   „ Urbanus, Sow.
   „ (Aturia) zic-zac, Bronn.
   Gasteropoda.
Actæon crenatus, Sow.
   „ simulatus, Sow.
Aporrhais Sowerbyi, Mant.
Bulimus ellipticus, Sow.
Bulla attenuata, Sow.
Cancellaria læviuscula, Sow.
Cassidaria nodosa, Brand.
   „ striata, Sow.
Cerithium Charlesworthii, Prest.
   „ Londinense, Edw. MS.
Chemnitzia sp.
Cypræa oviformis, Sow.
   „ Wetherelli, Edw.

Dentalium anceps, Sow.
Eulima subulata, Sow., not Mont.
Fusus bifasciatus, Sow.
   „ carinella, Sow.
   „ coniferous, Sow.
   „ curtus, Sow.
   „ interruptus, Sow.
   „ regularis, Sow.
   „ trilineatus, Sow.
   „ tuberosus, Sow.
Metula juncea, Sow.
Murex cristatus, Sow.
Natica labellata, Lam.
Nerita globosa, Sow.
Ovula (?) antiqua, Edw.
Phasianella (?).
Phorus extensus, Sow.
Pisania lineatissima, Edw. MS.
   „ dubia, Edw. MS.
Pleurotoma acuminata, Edw.
   „ denticula, Bast.
   „ helix, Edw.
   „ teretrium, Sow.
Pyrula angulata, Edw.
   „ Smithii, Sow.
   „ tricostata, Desh.
Rissoa (?).

Rostellaria lucidia, Sow.
Scallaria reticulata, Sow., not Brand.
Sigaretus canaliculatus, Lam.
Solarium patulum, Lam.
Triton fasciatus, Edw.
Turritella imbricataria, Lam.
   ,,   scalaroides, Sow. (?)
Typtus muticus, Sow.
Voluta elevata, Sow.
   ,,   protensa, Sow.
   ,,   tricorona, Sow.
   ,,   Wetherelli, Sow.

Lamellibranchiata.

Anomia tennistriata, Desh.
Arca impolita, Sow.
   ,,   nitens, Sow.
Astarte rugata, Sow.
Avicula arcuata, Sow.
   ,,   media, Sow.
   ,,   papyracea, Sow.
Cardium Plumsteadiense, Sow.
Corbula globosa, Sow.
Cryptodon angulatum, Sow.
   ,,   Goodhallii, Sow.
Cyprina planata, Sow.
Leda amygdaloides, Sow.
   ,,   minima, Sow.
   ,,   oblata, Wood.
Modiola elegans, Sow.
Neæra inflata, Sow.
Nucula Bowerbankii, Sow.
   ,,   Wetherelli, Sow.

Pecten duplicatus.
Pholadomya margaritacea, Sow.
Pinna affinis, Sow.
Protocardium nitens, Sow.
Syndosmya splendens, Sow.
Teredo antenautæ, Sow.
Verticordia sulcata, Sow.

Brachiopoda.

Terebratulina striatula, Sow.

Polyzoa.

Flustra crassa, Desm.
Nidulites sp.

Crustacea.

Archæocarabus Bowerbankii, M'Coy.
Dromilites Lamarckii, M'Coy.
Hoploparia Belli, M'Coy.
Thenops scyllariformis, Bell.
Trachysoma scabrum, Bell.
Xanthopsis Leachii, Desm.

Annelida.

Ditrupa plana, Sow.
Serpula prismatica, Sow.
   ,,   trilineata, Sow.
Vermicularia Bognoriensis, Mant.

Echinodermata.

Pentacrinus Oakshottianus, Forb.
   ,,   subbasaltiformis, Mill.

Actinozoa.

Dasmia Sowerbyi, M. Edw.
Graphularia Wetherelli, M. Edw.

FOSSILS from other Places in the Parish.

Gasteropoda.

Bulla sulcifera, Edw. MS. ; Hampstead
    Railway.
Cassidaria striata, Sow. ; Belsize.

Cassidaria sulcaria, Desh. ; Hampstead
    Tunnel.
Dentalium bisiphonatum, Edw. ; Hampstead Railway.

Dentalium subcaniculatum, Edw.; Hampstead Railway.

Litiopa primæva, Edw. ; Hampstead Railway.

Murex subcoronatus, d'Orb ; Kilburn.

Phorus extensus, Sow. ; Belsize Lane.

Pleurotoma fasciolata, Edw. ; Kilburn.

„ teretrium, Edw. ; Kilburn.

Pleurotoma tæniolata ; Kilburn.

Pyrula angulata, Edw. ; Belsize Lane.

Voluta nodosa, Sow. ; Kilburn.

Lamellibranchiata.

Nucula compressa, Sow. ; Hampstead Tunnel.

„ lamellata, Edw. MS. ; Hampstead Railway.

BASEMENT BED OF THE LONDON CLAY.

FOSSILS from LOWER HEATH. (Well sinking.)

Pisces.
Otodus teeth.
Fish scales.

Gasteropoda.
Aporrhais Sowerbyi, Mant.
Natica labellata, Lam.
Pleurotoma sp.

Lamellibranchiata.
Cardium Laytoni, Mor.
„ nitens, Sow.
Cytherea obliqua, Desh.
Nucula sp.
Panopæa intermedia, Sow.

In addition to the species included in the above Lists the following are recorded as from " Hampstead," locality not further specified.

Gasteropoda.
Aporrhais Margarini, De Kon., var. speciosa.
Bulla decollata, Edw. MS.
Cancellaria aveniformis, Edw. MS.
Cerithium suturale, Edw. MS.
Chrysodomus fasciatus, Edw. MS.
Euthria nanatiphora, Edw. MS.
Ficula subangulata, Edw. MS.
Murex subcristatus, d'Orb.
Pisania lineatissima, Edw. MS.
„ transversaria, Edw. MS.
Pleurotoma crassa, Edw.
„ helix, Edw., var. rinca.
„ symmetrica, Edw.

Pleurotoma terebralis, Lam., var. concinna, Edw.
„ terebralis, Lam., var. ditropis, Edw.
„ teretrium, Edw., var. crebrilinea, Edw.
„ teretrium, Edw., var. nanodis, Edw.
Scalaria cylindrica, Edw. MS.
„ cymæa, Edw. MS.
„ reticulata, Sol., var. depressa, Edw. MS.
Solarium pulchrum, Sow., var. primævum, Edw.

Lamellibranchiata.	Astarte filigera, Edw.
Astarte tenera, Morris, var. Hampsteadiensis, S. Wood.	Corbula pisum, Sow.
	Nucula consors, S. Wood.

BIBLIOGRAPHY OF THE GEOLOGY OF HAMPSTEAD.

Geological Survey of England and Wales:

Memoirs, vol. iv., The Geology of the London Basin. By William Whitaker, B.A.

Guide to the Geology of London and the Neighbourhood. By William Whitaker, B.A., F.G.S.

Map—Sheet, "London and its Environs."

Horizontal Section, Sheet 79.

Proceedings of the Geologists' Association:

On the Geology of Hampstead, Middlesex. By Caleb Evans. Vol. iii., p. 21.

On a New Section of the Upper Bed of the London Clay. By Caleb Evans, F.G.S. Vol. ii., p. 283.

Reports of Excursions to Hampstead. Vol. ii., p. 40; vol. iii., p. 67; vol. iv., p. 155; vol. v., p. 160; vol. x., p. 148. By Caleb Evans, F.G.S., and J. Logan Lobley, F.G.S.

Geological Society, Quarterly Journal:

On the Thickness of the London Clay; on the Relative Position of the Fossiliferous Beds of Sheppey, Highgate, Harwich, Newnham, Bognor, &c.; and on the Probable Occurrence of the Bagshot Sands in the Isle of Sheppey. By Joseph Prestwich, Jun., F.R.S., F.G.S., &c. Vol. x., p. 415.

On the Boring through the Chalk at Kentish Town. By Joseph Prestwich, F.R.S. Vol. xii., p. 124.

On the Pebble Beds of Middlesex, Essex, and Herts. By Searles V. Wood, F.G.S. Vol. xxiv., p. 464.

Proceedings of the Geological Society :
  On the Hydrographical Basin of the Thames. By the Rev. W.
    D. Conybeare. Vol. i., p. 145.
  Observations on the London Clay of the Highgate Archway.
    By N. T. Wetherell. Vol. i., p. 409.
  On Ophiura found at Child's Hill to the N.W. of Hampstead.
    By N. T. Wetherell.
Transactions of the Geological Society :
  Observations on a Well dug on the south side of Hampstead
    Heath. By N. T. Wetherell. Series 2, vol. v., p. 131.
Transactions of the Middlesex Natural History Society :
  The Geology of the Parish of Hampstead. By J. Logan Lobley,
    F.G.S., F.R.G.S., &c., 1886-7, p. 64.
Philosophical Transactions, 1792, p. 115 :
  Description of Kilburn Wells and Analysis of their Water. By
    J. S. Schmeisser.
London and Edinburgh Philosophical Magazine, series 3, vol. ix., p.
    462 :
  Observations on some of the Fossils of the London Clay, and
    in particular those Organic Remains which have been recently
    discovered in the Tunnel for the London and Birmingham Rail-
    road. By Nath. Thomas Wetherell, Esq., F.G.S., M.R.C.S.
Magazine of Natural History, series 2, vol. ii., p. 625 :
  Highgate Resin. By N. T. Wetherell.

The Ground Beneath Us. By Joseph Prestwich, London.
The Topography and Natural History of Hampstead. By J. J.
  Park (1818).
View of the Agriculture of Middlesex, 2nd ed. Account of Soils,
  Wells, &c. By J. Middleton (1813).
An Account of the Neutral Saline Waters recently discovered at
  Hampstead. By T. Goodwin (1801).

THE FLORA OF HAMPSTEAD.

By HENRY T. WHARTON, M.A., M.R.C.S., F.Z.S., &c.

In this list of the plants known to occur in the district, the authorities for the names are omitted, since the order and nomenclature of "The London Catalogue of British Plants" (8th ed., 1886) are followed throughout. Varieties are not noted, as they have never been sufficiently worked out. Hence one plant, *Ranunculus aquatilis*, stands under its old name. Trimen and Dyer's "Flora of Middlesex" (1869) has been freely used. Doubtful plants are passed over. Collectors extirpate rarities, but *Maianthemum* grows abundantly in a certain part of Caen Wood. There is still much left for conscientious investigators to add. Out of the 1,760 Phanerogams enumerated in "The London Catalogue," only 455 species appear in this list.

RANUNCULACEÆ :
  Anemone nemorosa.
  Ranunculus aquatilis (varieties un-
        determined).
      „   hederaceus.
      „   sceleratus.
      „   Flammula.
      „   auricomus.
      „   acris.
      „   repens.
      „   bulbosus.
      „   arvensis.
      „   Ficaria.
  Caltha palustris.

BERBERIDEÆ :
  Berberis vulgaris.
NYMPHÆACEÆ :
  Nuphar luteum.
  Nymphæa alba.
PAPAVERACEÆ :
  Papaver Rhœas.
  Chelidonium majus.
FUMARIACEÆ :
  Fumaria officinalis.
CRUCIFERÆ :
  Nasturtium officinale.
      „   sylvestre.
      „   palustre.

Barbarea vulgaris.
Cardamine amara.
    ,,   pratensis.
    ,,   hirsuta.
Erophila vulgaris.
Cochlearia Armoracia.
Sisymbrium officinale.
    ,,   Alliaria.
Brassica nigra.
    ,,   Sinapis.
    ,,   alba.
Capsella Bursa-pastoris.
Senebiera Coronopus.
Lepidium ruderale.
    ,,   campestre.
Raphanus Raphanistrum.
VIOLARIEÆ :
Viola palustris.
    ,, odorata.
    ,, sylvatica.
    ,, tricolor.
POLYGALEÆ :
Polygala vulgaris.
CARYOPHYLLEÆ :
Lychnis diurna.
    ,,   Flos-cuculi.
Cerastium quaternellum.
    ,,   semidecandrum.
    ,,   glomeratum.
    ,,   triviale.
Stellaria media.
    ,,   Holostea.
    ,,   graminea.
    ,,   uliginosa.
Arenaria trinervis.
    ,,   serpyllifolia.
Sagina procumbens.
Spergula arvensis.
Lepigonum rubrum.

PORTULACEÆ :
Montia fontana.
HYPERICINEÆ :
Hypericum perforatum.
    ,,   quadrangulum.
    ,,   humifusam.
    ,,   pulchrum.
MALVACEÆ :
Malva moschata.
    ,,   sylvestris.
    ,,   rotundifolia.
TILIACEÆ :
Tilia vulgaris.
LINEÆ :
Linum usitatissimum.
GERANIACEÆ :
Geranium molle.
    ,,   pusillum.
    ,,   dissectum.
    ,,   Robertianum.
Oxalis Acetosella.
ILICINEÆ :
Ilex Aquifolium.
CELASTRINEÆ :
Euonymus europæus.
RHAMNEÆ :
Rhamnus Frangula.
SAPINDACEÆ :
Acer Pseudo-platanus.
    ,, campestre.
LEGUMINOSÆ :
Genista anglica.
Ulex europæus.
    ,, nanus.
Cytisus scoparius.
Ononis repens.
Medicago lupulina.
    ,,   maculata.
Trifolium pratense.

Trifolium medium.
,, arvense.
,, repens.
,, procumbens.
Lotus corniculatus.
,, pilosus.
Ornithopus perpusillus.
Vicia hirsuta.
,, Cracca.
,, sepium.
Lathyrus Nissolia.
,, macrorrhizus.
ROSACEÆ :
Prunus communis.
,, Avium.
,, Padus.
Spiræa Ulmaria.
Rubus Idæus (varieties not distinguished).
Geum urbanum.
Fragaria vesca.
Potentilla Fragariastrum.
,, Tormentilla.
,, reptans.
,, Anserina.
Alchemilla arvensis.
Poterium Sanguisorba.
Rosa canina.
,, arvensis.
Pyrus torminalis.
,, Aria.
,, Aucuparia.
,, communis.
,, Malus.
Cratægus Oxyacantha.

SAXIFRAGEÆ :
Saxifraga tridactylites.
Chrysosplenium oppositifolium.

DROSERACEÆ :
Drosera rotundifolia.
HALORAGEÆ :
Callitriche vernalis.
,, hamulata.
LYTHRARIEÆ :
Lythrum Salicaria.
Peplis Portula.
ONAGRARIEÆ :
Epilobium hirsutum.
,, parviflorum.
,, montanum.
,, tetragonum.
,, obscurum.
,, palustre.
Circæa lutetiana.
CUCURBITACEÆ :
Bryonia dioica.
UMBELLIFERÆ :
Hydrocotyle vulgaris.
Sanicula europæa.
Apium nudiflorum.
,, inundatum.
Sison Amomum.
Sium erectum.
Ægopodium Podagraria.
Pimpinella Saxifraga.
Conopodium denudatum.
Anthriscus sylvestris.
Æthusa Cynapium.
Angelica sylvestris.
Heracleum Sphondylium.
Caucalis Anthriscus.
ARALIACEÆ :
Hedera Helix.
CORNACEÆ :
Cornus sanguinea.
CAPRIFOLIACEÆ :
Adoxa Moschatellina.

Sambucus nigra.
Viburnum Opulus.
„ Lantana.
Lonicera Periclymenum.
RUBIACEÆ :
Galium verum.
„ Mollugo.
„ saxatile.
„ palustre.
„ uliginosum.
„ Aparine.
Asperula odorata.
Sherardia arvensis.
VALERIANEÆ :
Valeriana dioica.
„ officinalis.
DIPSACEÆ :
Dipsacus sylvestris.
„ pilosus.
Scabiosa succisa.
„ arvensis.
COMPOSITÆ :
Eupatorium cannabinum.
Solidago Virgaurea.
Bellis perenne.
Gnaphalium uliginosum.
„ sylvaticum.
Bidens cernua.
„ tripartita.
Achillea Millefolium.
„ Ptarmica.
Anthemis nobilis.
Chrysanthemum Leucanthemum.
„ Parthenium.
Matricaria inodora.
„ Chamomilla.
Tussilago Farfara.
Senecio vulgaris.
„ sylvati

Senecio crucifolius.
„ Jacobæa.
„ aquaticus.
Arctium minus.
Carduus nutans.
Cnicus lanceolatus.
„ palustris.
„ arvensis.
Serratula tinctoria.
Centaurea nigra.
„ Cyanus.
Lapsana communis.
Picris hieracioides.
„ echioides.
Crepis virens.
Hieracium Pilosella.
„ murorum.
„ vulgatum.
„ umbellatum.
„ boreale.
Hypochæris radicata.
Leontodon hirtus.
„ hispidus.
„ autumnalis.
Taraxacum officinale.
Lactuca muralis.
Sonchus oleraceus.
„ asper.
„ arvensis.
Tragopogon pratensis.
CAMPANULACEÆ :
Jasione montana.
Campanula rotundifolia.
VACCINIACEÆ :
Vaccinium Myrtillus.
ERICACEÆ :
Calluna Erica.
Erica Tetralix.
„ cinerea.

PRIMULACEÆ :
Primula vulgaris.
„ veris.
Lysimachia vulgaris.
„ Nummularia.
„ nemorum.
Anagallis arvensis.
OLEACEÆ :
Fraxinus excelsior.
Ligustrum vulgare.
APOCYNACEÆ :
Vinca minor.
GENTIANEÆ :
Erythræa Centaurium.
Menyanthes trifoliata.
BORAGINEÆ :
Cynoglossum officinale.
Symphytum officinale.
Myosotis cæspitosa
„ palustris.
„ arvensis.
„ versicolor.
CONVOLVULACEÆ :
Calystegia Sepium.
SOLANACEÆ :
Solanum Dulcamara.
„ nigrum.
SCROPHULARINEÆ :
Verbascum nigrum.
Linaria Cymbalaria.
„ vulgaris.
Scrophularia aquatica.
„ nodosa.
Limosella aquatica.
Digitalis purpurea.
Veronica hederæfolia.
„ agrestis.
„ arvensis.
„ serpyllifolia.

Veronica officinalis.
„ Chamædeys.
„ montana.
„ scutellata.
„ Anagallis.
„ Beccabunga.
Euphrasia officinalis.
Bartsia Odontites.
Pedicularis palustris.
„ sylvatica.
Melampyrum pratense.
Rhinanthus Crista-galli.
VERBENACEÆ :
Verbena officinalis.
LABIATÆ :
Mentha hirsuta.
Lycopus europæus.
Thymus Serpyllum.
Nepeta Glechoma.
Scutellaria galericulata.
„ minor.
Prunella vulgaris.
Marrubium vulgare.
Stachys Betonica.
„ palustris.
„ sylvatica.
„ arvensis.
Galeopsis Tetrahit.
Lamium purpureum.
„ album.
„ Galeobdolon.
Ballota nigra.
Teucrium Scorodonia.
Ajuga reptans.
PLANTAGINEÆ :
Plantago major.
„ media.
„ lanceolata.
„ Coronopus.

CHENOPODIACEÆ:
Chenopodium polyspermum.
  ,, album.
  ,, ficifolium.
  ,, rubrum.
  ,, glaucum.
Atriplex patula.
  ,, hastata.

POLYGONACEÆ:
Polygonum Convolvulus.
  ,, aviculare.
  ,, Hydropiper.
  ,, Persicaria
  ,, lapathifolium.
  ,, amphibium.
  ,, Bistorta.
Rumex conglomeratus.
  ,, sanguineus.
  ,, palustris.
  ,, obtusifolius.
  ,, acutus.
  ,, crispus.
  ,, Hydrolapathum.
  ,, Acetosa.
  ,, Acetosella.

EUPHORBIACEÆ :
Euphorbia Helioscopia.
  ,, amygdaloides.
  ,, Peplus.
Mercurialis peremus.
  ,, annua.

URTICACEÆ:
Ulmus montana.
  ,, campestris.
Humulus Lupulus.
Urtica dioica.
  ,, ureus.
Parietaria officinalis.

CUPALIFERÆ :
Betula alba.
Alnus glutinosa.
Carpinus Betulus.
Corylus Avellana.
Quercus Robur.
Castanea sativa.
Fagus sylvatica.

SALICINEÆ :
Salix fragilis.
  ,, alba.
  ,, purpurea.
  ,, viminalis.
  ,, Smithiana.
  ,, cinerea.
  ,, aurita.
  ,, Caprea.
  ,, repeus.
Populus alba.
  ,, tremula.
  ,, nigra.

CERATOPHYLLEÆ:
Ceratophyllum demersum.

CONIFERÆ :
Taxus baccata.
Pinus sylvestris.

HYDROCHARIDEÆ :
Elodea canadensis.

ORCHIDEÆ :
Listera ovata.
Orchis Morio.
  ,, mascula.
  ,, maculata.

IRIDEÆ :
Iris Pseudacorus.
Crocus vernus.

AMARYLLIDEÆ :
Narcissus Pseudo-narcissus.

DIOSCOREÆ :
  Tamus communis.
LILIACEÆ :
  Ruscus aculeatus.
  Polygonatum multiflorum.
  Maianthemum Convallaria.
  Convallaria majalis.
  Allium ursinum.
  Scilla nutans.
  Paris quadrifolia.
JUNCACEÆ :
  Juncus bufonius.
    ,, glaucus.
    ,, effusus.
    ,, supinus.
    ,, lamprocarpus.
    ,, acutiflorus.
  Luzula pilosa.
    ,, maxima.
    ,, campestris.
    ,, multiflora.
TYPHACEÆ :
  Typha latifolia.
  Sparganium ramosum.
AROIDEÆ :
  Arum maculatum.
  Acorus Calamus.
LEMNACEÆ :
  Lemna trisulca.
    ,, minor.
    ,, gibba.
    ,, polyrrhiza.
ALISMACEÆ :
  Alisma Plantago.
  Butomus umbellatus.
NAIADACEÆ :
  Potamogeton natans.
    ,, polygonifolius.
    ,, perfoliatus.

Potamogeton crispus.
CYPERACEÆ :
  Scirpus setaceus.
    ,, sylvaticus.
  Eriophorum augustifolium.
  Carex pulicaris.
    ,, disticha.
    ,, paniculata.
    ,, vulpina.
    ,, muricata.
    ,, echinata.
    ,, remota.
    ,, ovalis.
    ,, Goodenowii.
    ,, panicea.
    ,, glauca.
    ,, pilulifer.
    ,, pendula.
    ,, sylvatica.
    ,, binervis.
    ,, flava.
    ,, hirta.
    ,, paludosa.
GRAMINEÆ :
  Phalaris canariensis.
    ,, arundinacea.
  Anthoxanthum odoratum.
  Alopecurus agrestis.
    ,, fulvus.
    ,, geniculatus.
    ,, pratensis.
  Milium effusum.
  Phleum pratense.
  Agrostis canina.
    ,, alba.
    ,, vulgaris.
  Aira carophyllea.
    ,, præcox.
  Deschampsia cæspitosa.

Deschampsia flexuosa.
Holcus mollis.
„ lanatus.
Trisetum flavescens.
Arrhenatherum avenaceum.
Sieglingia decumbeus.
Phragmites communis.
Cynosurus cristatus.
Catabrosa aquatica.
Melica uniflora.
Dactylis glomerata.
Briza media.
Poa annua.
„ nemoralis.
„ compressa.
„ pratensis.
„ trivialis.
Glyceria fluitaus

Glyceria plicata.
„ aquatica.
Festuca sciuroides.
„ ovina.
„ elatior.
Bromus giganteus.
„ asper.
„ sterilis.
„ racemosus.
„ mollis.
Brachypodium sylvaticum.
Lolium perenne.
„ temulentum.
Agropyron repens.
Nardus stricta.
Hordeum pratense.
„ murinum.

THE INSECT FAUNA OF HAMPSTEAD.

By the Rev. F. A. WALKER, D.D., F.I.S., F.E.S., &c.

The subjoined List of Insects occurring in the parish of Hampstead represents the result of careful observation during the last four years. Nearly all the species here specified can be verified by reference to my cabinet of insects taken by myself. Those indicated by an asterisk have not come under my own notice, but were observed and taken by Mr. John Cother Webb, an experienced entomologist and a careful observer.

I have found the hedgerow that skirts the Midland Railway in the field leading to Child's Hill a good locality for Hymenoptera and likewise Diptera. Hampstead Heath has been famed for its Hymenoptera. In former days, as a resident at Highgate, I was far better acquainted with the northern than the southern end of Hampstead, when Bishop's Wood and the palings skirting the Caen Wood estate in Hampstead Lane were favourite resorts of mine in pursuit of Entomology. It would be interesting to ascertain if *Vanessa polychloros*, which I have not found in South Hampstead, still occurs in Bishop's Wood as it did thirty years ago. *Cynthia* (or *Vanessa*) *Hampstediensis* was caught on Hampstead Heath ; but Kirby, in his general list of butterflies, gives *Cynthia Hampstediensis* as a synonym of *Junonia vellida*, which is a species found in Java, Sumatra, and Australia. According to this statement the insect in question would be an instance of a

6

foreign butterfly, accidentally shipped in the chrysalis state, having come to London with foreign produce, unless some one had imported pupæ and let the insects fly as soon as they emerged. I have, however, a dim recollection of hearing my father state that *C. Hampstediensis* was a North American species allied either to our Painted Lady or Small Tortoiseshell.

Much still remains to be done, particularly by any specialist who would work up the *Tineæ* among the Lepidoptera, and the more minute species among the Coleoptera.

LEPIDOPTERA, RHOPALOCERA (Butterflies).

Pieridæ.
Pieris brassicæ.
  „   rapæ.
  „   napi.
Anthocharis cardamines.
Gonepteryx rhamni.
Satyridæ.
Satyrus janira.
Cœnonympha pamphilus.

Lycænidæ.
Polyommatus Alexis.
Chrysophanus phlæas.
Vanessidæ.
Vanessa Io.
  „   Atalanta.
  „   cardui.
  „   urticæ.
Hesperidæ.
Pamphila Sylvanus.

LEPIDOPTERA HETEROCERA (Moths).

Sphingidæ.
Macroglossa Stellatarum.
Sphinx Ligustri (larva).
Smerithus Populi.
  „   Tiliæ.
Zygænidæ.
*Zygæna Filipendulæ.
Zeuzeridæ.
*Cossus ligniperda.
*Zeuzera Æsculi.
Cheloniidæ.
*Chelonia Caja.
*Arctia Menthastri.
*Arctia lubricipeda.

Bombycidæ.
*Bombyx neustria.
Hepialidæ (Swifts).
Hepialus lupulinus.
  „   humuli.
Liparidæ.
Liparis salicis.
Liparis auriflua.
Orgyia antiqua.
Cuspidates.
Pygæra bucephala.
Cerura vinula.
Cilix spinula.

LEPIDOPTERA HETEROCERA (Moths) *continued—*

Noctuæ.

*Leucania lithargyria.
   ,,   pallens.
*Hydrœcia nictitans.
*Xylophasia lithoxylea.
   ,,   polyodon.
*Mamestra persicariæ.
*Phlogophora meticulosa.
*Euplexia lucipara.
*Mania maura.
Catocala nupta.
Plusia gamma.
Acronycta psi.
Mamestra brassicæ.
Charœas graminis.
Tryphœna pronuba.
Apamea gemina.
Tapinostola fulva.
Noctua glareosa.
Tæniocampa stabilis.
Caradrina blanda.
Miana strigilis.
Acronycta megacephala.
Bryophila perla.
Apamea basilinea.
Euclidia mi.
Heliodes arbuti.

Geometers.

*Boarmia repandata.
Boarmia rhombordaria.
Camptogramma bilineata.
Crocallis elinguaria.
Abraxas grossulariata.

Halia wavaria.
Biston hirtaria.
Melanippe montanata
   ,,   fluctuata.
   ,,   subtristata.
Rumia cratægata.
Hemerophila abruptaria.
Chimatobia brumata.
*Uropteryx sambucata.
*Selenia illunaria.
*Ennomos angularia.
*Phigalia pilosaria.
*Amphydasis betularia.
*Anticlea derivata.
*Tanagra chœrophyllata.

Tortrices.

Tortrix heperana.
Penthina ochroleucana.

Tineina.

Hyponemeutidæ.

Hyponemeuta padellus.
Pterophorus pentadactylus.

Deltoides.

Hypena proboscidalis.

Pyralidæ.

Botys verticalis.
Pyralis fimbrialis.

Hydrocampidæ.

Cataclysta lemnata.

Crambidæ.

Crambus tristellus.
   ,,   patellus.

NEUROPTERA (Dragon-flies, &c.).

Dragon-flies {
  Æschnidæ.
  Æschna cyanea.
  Libellulidæ.
  Libellula depressa.
}

Dragon-flies {
  Agrionidæ.
  Agrion puella.
  ,,   elegans.
  Pyrrhosoma minium.
}

NEUROPTERA (Dragon-flies, &c.) *continued—*

Phryganidæ (Stone-flies).
Halesus stellatus.
Leptocerus niger.
Platipennia.
Sialis lutarius.
　Chrysopidæ (Golden-eyed flies).
Chrysopa phyllochroma.

Chrysopa septem-punctata.
　Panorpidæ (Scorpion-flies).
Panorpa germanica.
　　　„　communis.
Larvæ of Ephemeridæ.
　　„　Agrionidæ.
　　„　Phrygania.

HYMENOPTERA.

Bombus sylvarum.
　　„　senilis.
　　„　latreillelus.
　　„　subterraneus.
　　„　hortorum.
　　„　pratorum.
　　„　lapidarius.
　　„　derhamellus.
　　„　lapponicus.
　　„　lucorum.
　　„　terrestris.
　　„　muscorum.
Apathus vestalis (parasitic).
Megachile circumcincta.
　　　　„　centuncularis.
　　　　„　lignisceca.
　　　　„　willughbiella.
Osmia rufa.
　　„　ænea.
Andrena albicans.
　　„　nitida.
　　„　nigro-ænea.
　　„　xanthura.
Anthophora acervorum.
　Apidæ.
Apis mellifica (Hive-bee).

Humble-bees. *(bracket label)*
Other genera of wild bees. *(bracket label)*

Apis Ligustica (Ligurian hive-bee).
　Chrysididæ.
Chrysis ignita.
　Vespidæ.
Ground-wasps. { Vespa vulgaris.
　　　　　　　{　„　germanica.
Polynerus quadratus.
Nomada alternata (parasitic).
　Chalcidites (Parasites of pupa of
　　　　Vanessa urticæ).
　　　„　sp.
　Ichneumonidæ.
Ophion luteum.
Lampronota setosa.
　　　„　bellator.
　　　„　murina.
　　　„　sp.
Ichneumon flavoniger.
　Formicidæ.
Ants. { Formica nigra.
　　　{　„　umbrata.
　　　{　„　flava.
Trichiosoma vitellina.
Cryptus ornatus.
Selandria annulata.

DIPTERA.

Eristalis arbustorum.
　„　tenax.

Eristalis pertinax.
　„　intricarius.

DIPTERA *continued—*

Syrphidæ.
Syrphus pyrastri.
  ,,   luniger.
  ,,   ribesii.
  ,,   balteata.
  ,,   corollæ.
Helophilus pendulus.
  ,,   frutetorum.
  ,,   trivittatus.
Sphærophoria tæniata.
Hyetodena.
Tabanus autumnalis.
Platychirus clypeatus.
  ,,   fulviventris.
  ,,   albimana.
Bibio Marci (St. Mark's Fly).
  ,,   hortulanus.
Syritta pipiens.

Melanostoma mellina.
Bluebottle-flies { Musca (Lucilia) Cæsar.
Calliphora erythrocephala.
  ,,   greenlandica.
  ,,   vomitoria.
Hæmatopota pluvialis.
Polietes lardaria.
Olivieria lateralis.
Scatophaga stercoraria (Dung-fly).
Empis scutellata.
  ,,   livida.
  ,,   tessellata.
Culex pipiens (Common Gnat).
Ptychopteryx contaminata.
Leptis scolopaceus.
Sarcophaga sinuata.
  ,,   melanura.
  ,,   hæmorrhoidalis.

COLEOPTERA.

Harpalus æneus.
Clivina fossor.
Xantholinus linearis.
Quedius tristis.
Creophilus maxillosus.
Philoxuthus laminatus.
  ,,   æneus.
Aleochara fuscipes.
Philobius uniformis.
Malachius bimaculatus.
Agriotes linearis.
Bembidium.
Luthrobium fulvipenne.
Soldier-beetles { Telephorus bicolor.
  ,,   lividus.
  ,,   flavilabris.
  ,,   melanurus.

Telephorus flavilabris var.
  ,,   testaceus.
Cetonia aurata.
Coccinella bipunctata.
  ,,   septem-punctata (Lady-birds).
Carabus violaceus.
  ,,   nemoralis.
  ,,   monilis.
Mycetophaga 4—pustulata.
Crioceris asparagi.
Pterostichus madidus.
  ,,   vulgaris.
  ,,   cupreus.
Nebria brevicollis.
Argutus vernalis.
Argutus amara.

COLEOPTERA *continued—*

Pyrrhochroa rubens (Cardinal-beetle).
Stenus.
Ocypus olens.
Clytus mysticus.
Phytœcia cylindrica.
Hylobius abietis.
Balaninus turbatus.

Water-beetles.
Gyrinus natator.
Agabus bipunctatus.
Henophorus aquaticus.
Hydrobius fuscipes.
Acilius sulcatus.
Haliplus lineatocollis.
Hyphidrus ovatus.
Hydroporus planus.

HEMIPTERA (Bugs).

Tropicoris rufipes.
Pantilius tunicatus.
Ptyclus bifasciatus.
Tettigonia viridis.
Oncognathus chenopodii.

Water-bugs.
Notonecta glauca (Water Boatman).
　　　,,　　lutea.
Nepa grisea (Water Scorpion).
Gerris thoracica (Water Measurer).
Larva of Gerris.

ORTHOPTERA (Grasshopper, &c.).

Blatta orientalis (Cockroach).
Forficula auricularia (Earwig).

Pezotettix pedestris (Field Grass-hopper).

MYRIOPODA.

Lithobius forficatus.
Geophilus longicornis.

Julus.

ARACHNIDA.

Epeira diadema.
Phalangium longipes.
Drassus cupreus.

Tegenaria domestica.
Lycosa rapax.

ENTOMOSTRACA.

Porcellio scaber (Common Wood-louse).

Armadillo vulgaris (Pill or Armadillo Wood-louse).

THE BIRDS OF HAMPSTEAD.

By J. E. HARTING, F.L.S., F.Z.S.

IN spite of the fact that Hampstead now forms part of the great metropolis with which it is connected by long lines of houses, it still preserves enough of its rural character to attract many species of birds. This is due not only to the proximity of Lord Mansfield's woods and to the unenclosed portions of heath which, happily, still remain, but to the number of old gardens which contain shrubberies and trees of goodly size, with here and there an open lawn affording good feeding ground at early dawn for many a feathered visitor. Looking out at daybreak upon one of these open spaces, we have often been surprised at the number of Blackbirds, Thrushes, Starlings, Chaffinches, Robins, Hedge Sparrows, and Tits in sight at one time, while during the summer months it is no uncommon thing to find the shrubberies tenanted by Willow Wrens, Redstarts, Spotted Flycatchers, and other migratory birds which find a temporary home there. Even the Reed Warbler has been known to find its way into Hampstead gardens, and make its cup-shaped nest amongst the lilac bushes.

With these annual summer migrants rarer visitors now and then appear in the shape of a Ring Ouzel, or Golden Oriole, but they do not stay long. Finding no congenial spot in which to bring up their young, they pass on in search of haunts more suited to their respective habits, or, as unfortunately it too often happens, to meet the fate which invariably threatens all brightly-plumaged birds.

From the proximity of the Hampstead and Highgate woods numbers of small birds find their way into the gardens, at first visiting those on the outskirts of the town, and gradually pushing on as far as any vegetation appears to attract them, until the smoke-begrimed trees and the absence of insect life begin to warn them against further advance.

In the winter Fieldfares and Redwings appear in the garden trees and join the ranks of the resident Thrushes upon the lawns at daybreak, the Redwing being often mistaken for the Song Thrush by those whose eyes are unaccustomed to note the distinguishing characteristics of birds. Amongst the rarer winter visitors may be mentioned the Crossbill, which has several times been met with at Hampstead, usually during frost, frequenting the fir-trees to extract the seeds from the cones. Twenty years ago it was no uncommon thing to find both Snipe and Jack Snipe on the lower part of the Heath in winter, and such is their affection for old haunts that it would not surprise us even now to meet with one at daybreak on the ground they used to frequent. It would be useless to look for these and many other species there during the day, but they may still drop down as of yore in the twilight, and after resting all night and feeding in the morning betake themselves to quieter or less disturbed haunts.

In the following list an attempt has been made to give some idea of the great variety of bird life which may be noted in the neighbourhood of Hampstead, not, of course, in a single day, but by observation during the year, by those who are sufficiently acquainted with birds to detect them by their note, flight, or peculiar movements when on the ground, or amongst the trees. For this purpose it is desirable to distinguish those which are more or less resident throughout the year from those which are periodical visitors in summer or winter, and from those, again, of such rare occurrence that their advent can neither be prophesied nor explained.

Those who may desire to have fuller information than can be well afforded in these pages may be referred to a little book entitled "The Birds of Middlesex," in which Hampstead is frequently mentioned in connection with many interesting birds observed there.[1] The indication of this source of information, it is hoped, will excuse the brevity with which the subject is here treated.

RESIDENT SPECIES.

SPARROWHAWK, *Accipiter nisus.* Bred in Caen Wood in 1876.

KESTREL, *Falco tinnunculus.* Bred in Caen Wood in 1871 and 1873.

BARN OWL, *Strix flammea.* For some years a pair frequented an avenue, where they reared their young, in June, 1873. In 1880 one was seen. In 1881 none. In 1886 a pair again appeared.

TAWNY, OR BROWN OWL, *Strix aluco.* Nest found in 1869.

LONG-EARED OWL, *Otus vulgaris.* Once common in the Hampstead Woods. Nest in Caen Wood, 1871.

MISTLETOE THRUSH, *Turdus viscivorus.* Not uncommon.

SONG THRUSH, *Turdus musicus.* Nesting in the gardens.

BLACKBIRD, *Turdus merula.* Common. A pied one visited a garden on Haverstock Hill for three years in succession.

HEDGE SPARROW, *Accentor modularis.* In gardens and shrubberies.

REDBREAST, *Erithacus rubecula.* Generally a pair or two in most gardens of any size where there are large shrubs.

STONECHAT, *Saxicola rubicola.* On the Heath ; usually observed in summer, but a few stay the winter.

DARTFORD WARBLER, *Melizophilus undatus.* Seen on the Heath, October, 1870. A pair seen same place, May, 1872.

GOLDCREST, *Regulus cristatus.* May be looked for amongst larch and spruce fir. The nest has been found at Frognal.

GREAT TITMOUSE, *Parus major.* Common.

---

[1] "The Birds of Middlesex." By J. E. HARTING. 8vo, pp. 284. Published by Gurney and Jackson, Paternoster Row.

BLUE TITMOUSE, *P. cœruleus.* Common. Occasionally building in odd places. There were two nests one summer at Roslyn, Hampstead, one in a lamp-post, the other in a letter-box.

COAL TITMOUSE, *P. ater.* A pair nested for several years in a garden on Haverstock Hill.

MARSH TITMOUSE, *P. palustris.* To be distinguished from the last named by the absence of a white spot on the nape. Has been found in Caen Wood.

LONGTAILED TITMOUSE, *P. caudatus.* In the woods and thick hedgerows, and in similar haunts with the Goldcrest. Rarely builds near houses, though the late Mr. G. Daniel has recorded the discovery of a nest of this bird in a large willow tree in a garden at Bayswater (cf. Bennett's ed. White's "Selborne," p. 171, footnote).

PIED WAGTAIL, *Motacilla lugubris.* Common about the ponds on the Heath, and in large gardens where there are lawns. Nest found at Frognal. (See list of Summer Migrants.)

MEADOW PIPIT, *Anthus pratensis.* Mr. Mitford is of opinion that the majority leave for the breeding season, and are common again in autumn and winter, but in 1877 and 1878 he found pairs breeding on West Heath, Hampstead.

SKYLARK, *Alauda arvensis.* Not so common as formerly. During the winter months, when these birds flock, large numbers are netted by the London bird-catchers.

COMMON BUNTING, *Emberiza miliaria.* May be heard and seen in the fields below the Hampstead Woods.

BLACKHEADED BUNTING, *E. schœniclus.* In the summer of 1880 a pair nested on the wet, rushy part of the west side of Hampstead Heath.

CHAFFINCH, *Fringilla cœlebs.* A curiously-made nest covered with little bits of newspaper was found at Hampstead by Mr. Mitford, who also has some eggs of this bird of a pale blue colour, *without spots.*

HOUSE SPARROW, *Passer domesticus.* The commonest bird in Hampstead. For many years the young sparrows about the George Inn were remarked to be pied, but apparently assumed the normal colour at their autumnal moult.

TREE SPARROW, *Passer montanus.* In June, 1871, the late Mr. Blyth ("Zoophilus") obtained three half-fledged Tree Sparrows which were taken from a nest in the hole of a tree at Hampstead. (See *The Zoologist,* 1877, p. 24, and 1888, p. 355.)

HAWFINCH, *Coccothraustes vulgaris.* Frequently observed in Bishop's Wood and Caen Wood, where at one time it used to breed regularly, the nest being generally placed on the horizontal bough of an oak. It feeds on beech-mast, and on the seeds of the hornbeam, sycamore, and maple, and visits the gardens in search of green peas, of which the young birds are very fond.

GREENFINCH, *Coccothraustes chloris.* Not very numerous, but may be seen flocking with Sparrows in the winter, occasionally coming into the gardens in spring and summer.

GOLDFINCH, *Carduelis elegans.* Can hardly be called a resident now, though formerly no doubt it was so. For this the London bird-catchers are much to blame. It is now most often seen when on migration in little parties in the autumn.

SISKIN, *Carduelis spinus.* A flock was seen in autumn near Caen Wood, feeding on the seeds of an alder, and a pair had a nest there in the summer of 1853. On August 9, 1871, a bird-catcher took one without a call-bird in Caen Wood. It must be considered a spring and autumn visitor rather than a resident, and its nesting here is very exceptional.

LINNET, *Linota cannabina.* In a Linnet's nest on Hampstead Heath two eggs of the Cuckoo were found.

BULLFINCH, *Pyrrhula europæa.* The custom of laying the hedges has almost driven the Bullfinch away, but it may still occasionally be seen on the outskirts of the woods. A pair had a nest one year in a summer-house belonging to Lord Mansfield, and brought up the young.

STARLING, *Sturnus vulgaris.* Next to the Sparrow perhaps the commonest bird in the district, nesting under eaves and in holes of decayed trees. Numbers are netted in autumn by the London bird-catchers for the use of members of the gun clubs who profess to find sport in shooting them from traps.

ROOK, *Corvus frugilegus.* There were at one time three rookeries at Hampstead. In 1880 only two, and these much reduced in numbers; but a new one was commenced that year in the Hampstead Road at Chalk Farm. In 1880, near the rookery by St. Stephen's Church, a notice-board was put up on one of the trees stating that the adjoining house was to let. All the Rooks then deserted the place but one pair which remained and built a nest, but others came from the upper rookery and pulled the nest to pieces. In 1883 the old one at North End still flourished. In 1887 there were eight nests in the rookery at Chalk Farm.

CARRION CROW, *Corvus corone.* For more than fifty years a pair of these birds have bred at Mr. Basil Wood's at Hillfield, and were still there in 1887.

JACKDAW, *Corvus monedula.* Frequenting the church towers and old trees on the Roslyn Park Estate. Often seen in company with Rooks.

MAGPIE, *Pica rustica.* For some years there was a Magpie's nest in a tree that no boy could climb, by Mr. Gurney Hoare's house at the Grove, but the birds

were wantonly shot by a wretched bird-stuffer, and although the survivor twice
brought home a new mate, the bird-stuffer, who ought to have been pro-
secuted, killed them one by one, and none have been seen there now for
some years.

JAY, *Garrulus glandarius.* At one time used to breed regularly in Bishop's Wood;
but is now believed to be extinct there.

GREEN WOODPECKER, *Picus viridis.* Formerly not uncommon in Lord Mansfield's woods; now rarely seen in Caen Wood. In October, 1871, one was·
observed upon a tree near the Heath.

GREAT SPOTTED WOODPECKER, *Picus major.* Used to breed annually in Caen
Wood. Now a rare bird in the district. Observed at Squire's Mount in
May, 1867, and last observed at Hampstead in spring of 1868.

LESSER SPOTTED WOODPECKER, *Picus minor.* More often observed than any
other species of Woodpecker. Noted in Bishop's Wood, Caen Wood, The
Priory, Frognal, Roslyn Park, and in a large elm by St. Stephen's
Church.

TREE CREEPER, *Certhia familiaris.* Now and then seen in gardens at Hampstead. A very silent and undemonstrative little bird. To be looked for on old
and decayed trees, round the trunks of which it creeps stealthily in search of
lurking insects.

WREN, *Traglodytes Europœus.* In old gardens and shrubberies, and about wood-
stacks. A nest of this little bird in a garden on Haverstock Hill was built
in a hole in a wall and was not domed.

NUTHATCH, *Sitta cœsia.* One of the few birds which has not decreased in numbers
of late years. Not uncommon in spring and autumn where large trees abound,
particularly about North End.

KINGFISHER, *Alcedo ispida.* Has been observed on the Regent's Canal, in the
Botanical Gardens, and other places close to Hampstead, and in May, 1872,
a nest was found in the bank of a pond near Caen Wood.

RINGDOVE, OR WOODPIGEON, *Columba palumbus.* A few pairs generally build in
Lord Mansfield's woods. At one time quite common in Bishop's Wood.

STOCK DOVE, *Columba œnas.* Comes in flocks to Caen Wood in autumn, when
the beech-mast is plentiful.

PHEASANT, *Phasianus torquatus.* Still found in Bishop's Wood.

PARTRIDGE, *Perdix cinerea.* Generally half a dozen coveys or so are reared in the
fields below Bishop's and Turner's Woods, and in September, 1868, a brace
were shot early in the morning on the Heath !

PEEWIT, OR LAPWING, *Vanellus cristalus.* Breeds at no great distance from
Hampstead, but can hardly be regarded as more than a passing visitor.

MOORHEN, *Gallinula chloropus*. Now and then one in the fields below Hampstead, and in Lord Mansfield's ponds.

SUMMER MIGRANTS.

RED-BACKED SHRIKE, *Lanius collurio*. On the outskirts of the woods, and in thick hedgerows.

SPOTTED FLYCATCHER, *Muscicapa grisola*. In some years very common. On park palings and lower branches of trees, darting into the air after passing insects.

PIED FLYCATCHER, *Muscicapa atricapilla*. Rare. There were two pairs in Bishop's Wood, in May, 1859. Nest found in 1866. Bird seen in May, 1867. Four young in August, 1868.

RING OUZEL, *Turdus torquatus*. Observed feeding on the berries of the mountain ash in a garden on Haverstock Hill, in September, 1865. Seen also at Caen Wood.

REDSTART, *Ruticilla phœnicurus*. In some years common.

WHINCHAT, *Saxicola rubetra*. On the Heath.   '

WHEATEAR, *Saxicola œnanthe*. Lower Heath and gravel-pits. The late Mr. G. Daniel (many of whose observations are quoted in the footnotes to Bennett's Edition of White's "Selborne") had a Wheatear which was taken at Hampstead on February 14, 1834. Unless this bird had remained throughout the winter, as some affirm that it does (*e.g.*, Jesse, in a footnote to his edition of White's "Selborne"), this is the earliest date for its appearance which has been noted.

GRASSHOPPER WARBLER, *Acrocephalus locustella*. On the Heath, and breeding in Bishop's Wood. In July, 1866, a nest was cut out of mowing grass.

SEDGE WARBLER, *A. phragmitis*. Caen Wood pond.

REED WARBLER, *A. streperus*. Nesting in gardens—in lilac bushes—far from water.

NIGHTINGALE, *Sylvia luscinia*. Common every year in Bishop's Wood and Caen Wood. A great number used to be caught every spring at Highgate by London bird-catchers.

BLACKCAP, *S. atricapilla*. Common in Bishop's Wood and Caen Wood, and visiting gardens at Frognal and Belsize Park. Feeds much on ivy-berries.

GARDEN WARBLER, *S. hortensis*. In some years very common around Hampstead. A fine songster. Will sit at evening in a high oak-tree or thick bush, like a Nightingale, and maintain a continued warble for ten minutes without a pause.

WHITETHROAT, *Sylvia cinerea.* } Seen regularly every summer in woods and
LESSER WHITETHROAT, *S. curruca.* } hedgerows.

WOOD WREN, *Phylloscopus sibilatrix.* In woods only, nesting on the ground.

WILLOW WREN, *P. trochilus.* } In woods and gardens, the former species being the
CHIFF-CHAFF, *P. rufa.* } commoner of the two. A pure white Willow
} Wren, in the collection of Mr. F. Bond, was
} taken some years ago in the Highgate Woods.

WHITE WAGTAIL, *Motacilla alba.* In April, 1866, a pair observed at the pond on
the Heath. One, in the collection of Mr. F. Bond, was procured at Hampstead in the summer of 1868.

YELLOW WAGTAIL, *M. raii.* Annual summer visitor, frequenting the Heath and
open fields, especially where there are sheep; also on railway embankments.
(See list of Winter Visitors.)

TREE PIPIT, *Anthus arboreus.* Common in summer. Breeds in the more open
parts of Bishop's and Turner's Woods.

WRYNECK, *Jynx torquilla.* Heard commonly about the Hampstead Woods at the
time of its arrival in April.

CUCKOO, *Cuculus canorus.* An annual summer visitor to the Hampstead Woods.
The eggs have been found on the Heath in the nest of a Linnet. On
September 29, 1872, a young Cuckoo, full fledged, was caught by a cat in a
garden in the Adelaide Road, and was taken to Mr. F. Bond, who then
resided there, and in whose collection it is now preserved.

SWALLOW, *Hirundo rustica.* } Nesting every year in suitable localities.
MARTIN, *Chelidon urbica.* }

SAND MARTIN, *Cotile riparia.* Used formerly to breed regularly on the property
of Lord Mansfield at Caen Wood Farm, where an old sand bank was completely riddled with their holes, but the bank slipped and they deserted the
place. In June, 1863, and again in 1871, some eggs were taken from holes
in a pit-bank on the Heath.

SWIFT, *Cypselus apus.* Seen flying round the church steeples and high over the
houses, but not ascertained to breed anywhere in the Hampstead district.

NIGHTJAR, *Caprimulgus Europœus.* In the summer of 1852 the eggs were found
in Caen Wood, and this bird used to breed annually in Bishop's Wood, the
eggs being often placed at no great distance from the main road.

TURTLE DOVE, *Turtur auritus.* A regular summer visitor, a few pairs breeding in
the woods. There was a nest in Bishop's Wood in 1887.

QUAIL, *Coturnix vulgaris.* In the summer of 1866 two broods of Quail were
reared in the fields below Hampstead. In 1870 three nests of eggs were
mown out in a large grass field between Highgate Hill and Pond Street

Station. In 1871 a few birds were heard calling; in 1872 none. (See *Zoologist*, 1871, p. 2728.) A Quail was heard calling below West Heath on June 14, 1877. In the summer of 1881 their unmistakable note was again heard in the fields adjoining Parliament Hill, where in the following year a brood was hatched and the empty egg-shells discovered.

LANDRAIL, *Crex pratensis*. In the fields just referred to the Landrail used to breed regularly every summer, and perhaps does so still.

COMMON SANDPIPER, *Totanus hypoleucus*. During the period of its migration in spring and autumn this bird may be looked for in the early morning along the banks of ponds and streams, particularly on the Brent, within an easy walk from Hampstead. At the same season of the year many other wading birds are heard passing overhead, such as the Green Sandpiper, Redshank, and Ringed Plover, which regularly visit Kingsbury Reservoir and its tributary streams.

GULLS AND TERNS of various species pass about the same time, but at too great a distance to be named with certainty.

WINTER VISITORS.

SHORT-EARED OWL, *Otus accipitrinus*. On the Lower Heath, rough fields, and rushy ground.

FIELDFARE, *Turdus pilaris*. ⎱ In the fields in open weather; coming into the gardens
REDWING, *T. iliacus*. ⎰ during frost and snow.

GREY WAGTAIL, *Motacilla sulphurea*. An uncertain winter visitor. Has been seen at the Heath pond.

SNOW BUNTING, *Emberiza nivalis*. In November, 1871, a flock of fifty or sixty "Snow-flakes" stayed for some time about the Leg of Mutton Pond on Hampstead Heath.

BRAMBLING, *Fringilla montifringella*. Appears towards the end of autumn in small flocks. Caen Wood used to be a favourite spot for them, and numbers are sometimes taken by the London bird-catchers. Looks like a Chaffinch on the wing, but always distinguishable by the white rump.

HOODED CROW, *Corvus cornix*. Has been observed on at least two occasions at Hampstead in winter.

WOODCOCK, *Scolopax rusticula*. About November a few drop into the Hampstead Woods, where, one summer, a nest of the Willow Wren was found lined with a few Woodcock's feathers, showing that the winter visitor had remained until the arrival of a summer guest, and had possibly nested there too.

SNIPE AND JACK SNIPE occasionally visit the Lower Heath.

PEREGRINE FALCON, *Falco peregrinus*. Winter, 1857.

HOBBY, *F. subbuteo*. Bishop's Wood, 1863; Hampstead Heath, August, 1864, May, 1869, and May, 1872.

MERLIN, *F. a'salon*. Hampstead Heath, winter, 1857.

GOSHAWK, *Astur palumbarius*. A young male captured at Hampstead on September 3, 1872, was preserved by Burton, of Wardour Street.

GREY SHRIKE, *Lanius excubitor*. Occasionally in winter. Sometimes caught by the bird-catchers when striking at their "call-birds."

GOLDEN ORIOLE, *Oriolus galbula*. Seen at Frognal and Sandfield Lodge, during May, 1862.

BLACK REDSTART, *Ruticilla tithys*. Caught at Hampstead by C. Davy, bird-catcher, April 14, 1869, identified by E. Blyth and R. Mitford. In the autumn of 1868 a pair seen in Culverhouse's brickfield; female caught and deposited in the Zoological Gardens.

WAXWING, *Ampelis garrulus*. In November, 1866, there was a remarkable immigration of this species. Three were shot out of a flock on Hampstead Heath. Seven were seen and one shot at Highgate in December, 1866, and others were seen at Hampstead, in January, 1867.

ROCK PIPIT, *Anthus obscurus*. Rarely found inland, except during the period of its migration in spring and autumn. One, in the collection of Mr. R. Mitford, was taken by a bird-catcher on Hampstead Heath in November, 1871.

RICHARD'S PIPIT, *Anthus richardi*. The specimen of this bird, from which the coloured figure in Gould's "Birds of Great Britain" was drawn, was caught in a net at Highgate, October 4, 1866, and between that date and the end of November, 1866, four more were taken in the same neighbourhood. In October, 1869, three more (all immature) were caught at Hampstead by bird-catchers in the employ of Mr. Davy, of Kentish Town. Two of these are preserved in the collection of Mr. F. Bond.

CIRL BUNTING, *Emberiza cirlus*. Seen on Hampstead Heath in April, 1855, April, 1860, and June, 1872. It has several times been met with at Hendon and Kingsbury. The call-note of the male resembles that of the Lesser Whitethroat.

MEALY REDPOLL, *Linota linaria*. In the autumn of 1866 a few were met with on the Heath.

LESSER REDPOLL, *L. rufescens*. Occasionally seen in autumn hanging about the birch-trees. In October, 1887, several little flocks were observed.

TWITE, *L. flavirostris.* In October, 1866, a few of these birds were observed about the Heath, in company with Linnets.

SCARLET GROSBEAK, *Pyrrhula erythrina.* A hen bird of this rare species was taken at Caen Wood on October 5, 1870 (*fide* Bond, *Zoologist,* 1870, p. 2383; see also Yarrell's "British Birds," 4th ed., ii. p. 173).

CROSSBILL, *Loxia curvirostra.* A rare and uncertain visitor. Generally in small flocks in winter. Observed at Hampstead in Scotch fir-trees in 1855, 1859, 1868, and 1880. In February, 1880, a little flock visited the fir-trees by the Spaniards.

HOOPOE, *Upupa epops.* In the summer of 1830 one was shot near Caen Wood. In the spring of 1866 one was observed on the embankment of the Kentish Town Railway, and two were shot in 1875 in Gospel Oak Field near Hampstead.

The absence of any considerable pools, and the distance from any river, save the Brent in the adjoining parish of Hendon, sufficiently explain the scarcity of water-fowl in the district under notice. All that can be claimed or expected at the present day is the occasional glimpse of a passing flock travelling high overhead *en route* for haunts more congenial to their habits.

POSTSCRIPT.

The foregoing Lists bear witness to the great amount of Natural History interest that may be found in the metropolitan area ; and a very striking illustration of there being no lack of avian visitors to the immediate neighbourhood of London homes is furnished by the following List of Birds that have been seen in the Royal Botanic Gardens, Regent's Park—an area much nearer to the centre of the metropolis even than Hampstead—to which my attention was very kindly called by Mr. William Sowerby, F.L.S., the secretary of the Royal Botanic Society.

LIST OF BIRDS OBSERVED IN THE GARDENS OF THE ROYAL BOTANIC SOCIETY, REGENT'S PARK.

(From No. 4 of the Quarterly Record of the Society.)

Blackbird.	Linnet.	Thrush, Song.
Blackcap.	Nightingale.	„ Missel.
Bullfinch.	Partridge.	Tit, Great.
Chaffinch.	Quail.	„ Cole.
Chiffchaff.	Redpole.	„ Blue.
Cuckoo.	Redwing.	Tree Creeper.
Fieldfare.	Reed Warbler.	Water-wagtail.
Flycatcher.	Robin.	Wheatear.
Garden Warbler.	Rook.	Whitethroat.
Goldfinch.	Siskin.	„ Lesser.
Greenfinch.	Sparrow, House.	Wren, Gold-crested.
Jackdaw.	„ Tree.	„ Willow
Kingfisher.	„ Hedge.	„ Wood
Lark, Sky.	Starling.	Yellowhammer.
„ Pipit.	Swallow.	

I have great pleasure in adding the following interesting communication received for publication after the above was in type, from the illustrious and venerable Professor Sir Richard Owen, K.C.B., respecting the birds visiting the grounds of his residence.

SHEEN LODGE, RICHMOND PARK.
August 31, 1889.

MY DEAR PROFESSOR,—

*   *   *   *   *   *

Since it may be of interest to the readers of your forthcoming work, 'Hampstead Hill,' as showing the variety of avian species found in the neighbourhood of London, I may mention that my friend, the late Mr. Gould, our most distinguished ornithologist, was a frequent visitor; and on one special occasion, accompanied by a common friend, Mr. Broderip, who made zoology his chief recreation after the dry labours of the Bench, devoted a May-day to making a list of the birds that visited the garden. I made a list of the species, and communicated it to the weekly journal called *The Garden*. The list contains upwards of sixty species.

*   *   *   *   *

Sincerely yours,
RICHARD OWEN.

PROF. LOGAN LOBLEY, F.G.S.

Just before going to press the following note reached me from Dr. Henry Wharton :—

ADDENDUM TO THE FLORA OF HAMPSTEAD.

Solanaceae :—Atropa Belladonna.

A single shrub of Deadly Nightshade, of exceptional size, grows by the side of the Midland Railway a quarter of a mile north of Mill Lane. Some workmen, on Sept. 22, 1889, found its berries so sweet and palatable that they ate large quantities, and gave many to their friends, some of whom made a pie of them. In all, ten people, adults and children, were taken to St. Mary's Hospital, most of them half-blind and in a violent delirium ; but they all recovered. The discoverer ate quite half a pint, but subsequent vomiting saved him from the worst consequences. The plant must have been brought by the railway from some calcareous district, for it occurs nowhere else in Middlesex nearer than Harefield.

UNWIN BROTHERS, THE GRESHAM PRESS, CHILWORTH AND LONDON

WILLIAM WESLEY AND SON,

Scientific Booksellers and Publishers,

28, ESSEX STREET, STRAND, LONDON.

The following recently published circulars include a portion of their stock :—

NATURAL HISTORY AND SCIENTIFIC BOOK CIRCULAR; No. 90.—*ASTRO-NOMY AND MATHEMATICS.* (Containing 1,600 works.) Contents: Astronomy : Practical and Historical ; Ancient and Mythical ; Biographies ; Photography ; Plurality of Worlds ; Spectrum Analysis and Light. The Sun, the Moon, Comets and Meteors. Star Catalogues and Charts. Mathematical and Astronomical Tables : Geodesy and Navigation : Figure of the Earth ; Logarithms. Mathematics. Works on Astronomical Instruments. Recent Foreign and English Publications on Photography. Price 6d.

NATURAL HISTORY AND SCIENTIFIC BOOK CIRCULAR; No. 91.—*CRYP-TOGAMIC BOTANY.* (Containing 450 works.) Desmids and Diatoms, Algæ, Characeæ and Confervæ, Lichens, Fungi and Micro-Fungi, Musci and Hepaticæ, Ferns, General Crypto-gamic Botany. Price 3d.

NATURAL HISTORY AND SCIENTIFIC BOOK CIRCULAR; No. 92.—*BOTANY.* (Containing 2,000 works.) A fine collection of Botanical Portraits and Valuable and Impor-tant Works (new and second-hand) on Flora of Great Britain and Ireland, and European Countries—Flora of Africa, America, Asia, Australasia—Botanical Gardens—Pre-Linnæan Botany, Herbals, and Medical Botany—Phanerogamic Botany, including Orchids—Botanical Dictionaries—Economic and Practical Botany—Physiology of Plants—Botanical Microscopy. Price 6d.

NATURAL HISTORY AND SCIENTIFIC BOOK CIRCULAR; No. 93.—*MICRO-SCOPIC ZOOLOGY, ENTOMOLOGY, CONCHOLOGY.* (Containing over 1,400 works.) Price 1s. (Nearly out of print.)

Just Published.

NATURAL HISTORY AND SCIENTIFIC BOOK CIRCULAR; No. 94.—*PHYSICAL SCIENCE.* Contents : Meteorology, Electricity, Galvanism, Magnetism, Light, Optics, Microscopy, Photography, Heat, Sound, Natural Philosophy, Chemistry. 1,400 Works. Price 6d.

NATURAL HISTORY AND SCIENTIFIC BOOK CIRCULAR; No. 95.—*GEOLOGY AND MINERALOGY;* including the first part of the Library of the late *W. H. BAILY,* of the Geological Survey of Ireland. Contents : Geology (Practical, Historical, and General) ; Geological Maps and Diagrams ; Glaciers and Glacial Theory ; Petrography ; Volcanoes and Earthquakes ; Geology of Great Britain, Ireland, and Europe ; Geology of Africa, America, Asia, and Australasia ; Mineralogy ; Mines and Crystallography ; Coal, Gas, Gems, Gold, Meteorites. New Purchases. 1,500 Works. Price 6d.

NATURAL HISTORY AND SCIENTIFIC BOOK CIRCULAR; No. 96.—*PA-LAEONTOLOGY;* including the second part of the Library of the late *W. H. BAILY,* Palaeontologist to the Geological Survey of Ireland. About 1,000 Works. Price 4d.

NATURAL HISTORY AND SCIENTIFIC BOOK CIRCULAR; No. 97.—Contents : *ICHTHYOLOGY;* Reptilia and Amphibia ; General Zoology, including Ancient Works, Biographies, Classification, Darwinism, Manuals, Periodicals, Transactions of Societies ; Anatomy, Physiology, and Embryology. Price 4d.

In Preparation.

NATURAL HISTORY AND SCIENTIFIC BOOK CIRCULAR; No. 98.—Contents : Mammalia ; including Cetacea, Sirenia and Pinnipedia. Ornithology ; including Eggs and Nests. Faunas of Britain, the Continents of Europe, Africa, America, Asia, Australasia. Zoological Voyages and Geographical Works. Price 6d.

W. WESLEY AND SON, 28, ESSEX STREET, STRAND, LONDON.

BOOKS FOR NATURE-LOVERS.

A SELECTED LIST

OF

POPULAR SCIENTIFIC WORKS

PUBLISHED BY

ROPER AND DROWLEY,

11, LUDGATE HILL, LONDON, E.C.

. *Detailed Prospectuses and Press Opinions Post Free on application.*

By J. ELLARD GORE, F.R.A.S., &c., &c.

PLANETARY AND STELLAR STUDIES. With numerous very beautiful Illustrations from Photographs and Drawings by well-known Fellows of the Royal Astronomical Society. Crown 8vo, cloth extra, price 7s. 6d.

THE SCENERY OF THE HEAVENS: A Popular Account of Astronomical Wonders. With many very beautiful Illustrations, Photographs, &c., of Star Clusters and Nebulæ, from the Original Photographs taken at the Paris Observatory, and by Mr. ROBERTS at Liverpool, and Drawings from recent sketches by well-known Astronomers. Crown 8vo, cloth extra, price 10s. 6d.

By G. S. BOULGER, F.L.S., F.G.S., &c., Prof. of Botany at the City of London College, Author of "Familiar Trees," &c.

THE USES OF PLANTS: A Manual of Economic Botany. With Special References to Vegetable Products Introduced during the last Fifty Years. This work aims at giving a concise enumeration, with a Systematic Index of the Vegetable Substances in use in England as Food, Materia Medica, Oils, Gums, Rubbers, Dyeing, Tanning, and Paper-making Materials, Fibres, Timber, &c., both home-grown and imported, together with Short Essays on the recent progress of Vegetable Technology in its various branches. Crown 8vo, cloth extra, price 6s.

By "A QUEKETT CLUB MAN."

THE STUDENT'S HANDBOOK TO THE MICROSCOPE: A Practical Guide to its Selection and Management. In crown 8vo, cloth gilt, with 38 Illustrations, price 2s. 6d. ; half morocco, 4s. 6d.

MY MICROSCOPE AND SOME OBJECTS FROM MY CABI-NET: A Simple Introduction to the Study of the Infinitely Little. New edition, enlarged, with Nine Illustrations, and beautifully bound, cloth extra, gilt, price 2s. 6d.

MY TELESCOPE AND SOME OBJECTS IT SHOWS ME: A Simple Introduction to the Glories of the Heavens. Companion Volume to above, with Ten Illustrations, cloth extra, silver, price 2s. 6d.

By T. CHARTERS WHITE, M.R.C.S., L.D.S., F.R.M.S.

A MANUAL OF ELEMENTARY MICROSCOPICAL MANIPU-LATION, for the use of Amateurs. In foolscap 8vo, Illustrated, cloth gilt, price 2s. 6d.

LONDON: ROPER AND DROWLEY, 11, LUDGATE HILL, E.C.

By J. W. WILLIAMS, M.A., D.Sc.

THE SHELL COLLECTOR'S HANDBOOK FOR THE FIELD.

Full detailed Prospectus on application. In foolscap 8vo, limp cloth gilt, price 5s., with Fourteen Illustrations, and Interleaved for Notes.

By J. LOGAN LOBLEY, F.G.S., &c., Prof. of Physiography and Astronomy at the City of London College.

MOUNT VESUVIUS: A Descriptive, Historical, and Geological

Account of the Volcano and its Surroundings. With Maps and Illustrations on plate paper. In demy 8vo, cloth extra, gilt, price 12s. 6d.

HAMPSTEAD HILL, with Chapters on the Flora of Hampstead,

by H. T. WHARTON, M.A., M.R.C.S., F.Z.S., &c.; The Insect Fauna of Hampstead, by Rev. Dr. WALKER, F.L.S., F.R.G.S., &c.; and The Birds of Hampstead, by J. EDMUND HARTING, F.L.S., &c., &c. In foolscap 4to, with Map and full-page plates illustrating local scenery, cloth extra, gilt, price 2s. 6d.

GEOLOGY FOR ALL. With Tables of the Principal Rock-forming

Minerals, Geological Strata, and of the Approximate Thickness of Strata. In crown 8vo, cloth extra, price 2s. 6d.

List of Devotional Books published by ROPER & DROWLEY, post free.

List of recent Works of Fiction published by ROPER & DROWLEY, post free.

LONDON: ROPER AND DROWLEY, 11, LUDGATE HILL, E.C.

DARLINGTON'S NATURALIST SERIES.

Crown 8vo. Sixpence each.

The Birds, Wild Flowers, Ferns, Grasses and Mosses of the Vale of Llangollen.
The Birds, Wild Flowers, Ferns, Grasses and Mosses of the North Wales Coast.
The Birds, Wild Flowers, Ferns, Grasses and Mosses of the Rhyl and the Vale of Clwyd.
The Birds, Wild Flowers, Ferns, Grasses and Mosses of Bettws-y-coed.
The Birds, Wild Flowers, Ferns, Mosses and Grasses of Aberystwith and Cardigan Bay.

Crown 8vo. Fourpence each.

The Wild Flowers of the Vale of Llangollen.
The Wild Flowers of the North Wales Coast.
The Wild Flowers of Rhyl and the Vale of Clwyd.
The Wild Flowers of Bettws-y-coed.
The Wild Flowers of Aberystwith and Cardigan Bay.

. The Botanical and English name is given of each flower, together with the soil or situation, colour, growth, duration, and time of flowering.

Crown 8vo. Twopence each.

The Birds of the Vale of Llangollen.
The Birds of the North Wales Coast.
The Birds of Rhyl and the Vale of Clwyd.
The Birds of Bettws-y-coed.
The Birds of Aberystwith and Cardigan Bay.

LLANGOLLEN: LONDON:
RALPH DARLINGTON.   ROPER AND DROWLEY.